Selected Titles in This Series

734 **Michel Van den Bergh,** Blowing up of non-commutative smooth surfaces, 2001
733 **Milé Krajčevski,** Tilings of the plane, hyperbolic groups and small cancellation conditions, 2001
732 **Jan O. Kleppe, Juan C. Migliore, Rosa Miró-Roig, Uwe Nagel, and Chris Peterson,** Gorenstein liaison, complete intersection liaison invariants and unobstructedness, 2001
731 **Jesús Bastero, Mario Milman, and Francisco J. Ruiz,** On the connection between weighted norm inequalities, commutators and real interpolation, 2001
730 **Suhyoung Choi,** The decomposition and classification of radiant affine 3-manifolds, 2001
729 **Michael Grosser, Eva Farkas, Michael Kunzinger, and Roland Steinbauer,** On the foundations of nonlinear generalized functions I and II, 2001
728 **Laura Smithies,** Equivariant analytic localization of group representations, 2001
727 **Anthony D. Blaom,** A geometric setting for Hamiltonian perturbation theory, 2001
726 **Victor L. Shapiro,** Singular quasilinearity and higher eigenvalues, 2001
725 **Jean-Pierre Rosay and Edgar Lee Stout,** Strong boundary values, analytic functionals, and nonlinear Paley-Wiener theory, 2001
724 **Lisa Carbone,** Non-uniform lattices on uniform trees, 2001
723 **Deborah M. King and John B. Strantzen,** Maximum entropy of cycles of even period, 2001
722 **Hernán Cendra, Jerrold E. Marsden, and Tudor S. Ratiu,** Lagrangian reduction by stages, 2001
721 **Ingrid C. Bauer,** Surfaces with $K^2 = 7$ and $p_g = 4$, 2001
720 **Palle E. T. Jorgensen,** Ruelle operators: Functions which are harmonic with respect to a transfer operator, 2001
719 **Steve Hofmann and John L. Lewis,** The Dirichlet problem for parabolic operators with singular drift terms, 2001
718 **Bernhard Lani-Wayda,** Wandering solutions of delay equations with sine-like feedback, 2001
717 **Ron Brown,** Frobenius groups and classical maximal orders, 2001
716 **John H. Palmieri,** Stable homotopy over the Steenrod algebra, 2001
715 **W. N. Everitt and L. Markus,** Multi-interval linear ordinary boundary value problems and complex symplectic algebra, 2001
714 **Earl Berkson, Jean Bourgain, and Aleksander Pełczynski,** Canonical Sobolev projections of weak type $(1,1)$, 2001
713 **Dorina Mitrea, Marius Mitrea, and Michael Taylor,** Layer potentials, the Hodge Laplacian, and global boundary problems in nonsmooth Riemannian manifolds, 2001
712 **Raúl E. Curto and Woo Young Lee,** Joint hyponormality of Toeplitz pairs, 2001
711 **V. G. Kac, C. Martinez, and E. Zelmanov,** Graded simple Jordan superalgebras of growth one, 2001
710 **Brian Marcus and Selim Tuncel,** Resolving Markov chains onto Bernoulli shifts via positive polynomials, 2001
709 **B. V. Rajarama Bhat,** Cocylces of CCR flows, 2001
708 **William M. Kantor and Ákos Seress,** Black box classical groups, 2001
707 **Henning Krause,** The spectrum of a module category, 2001
706 **Jonathan Brundan, Richard Dipper, and Alexander Kleshchev,** Quantum Linear groups and representations of $GL_n(\mathbb{F}_q)$, 2001
705 **I. Moerdijk and J. J. C. Vermeulen,** Proper maps of toposes, 2000
704 **Jeff Hooper, Victor Snaith, and Min van Tran,** The second Chinburg conjecture for quaternion fields, 2000

(Continued in the back of this publication)

Blowing Up of Non-Commutative Smooth Surfaces

MEMOIRS
of the
American Mathematical Society

Number 734

Blowing Up of Non-Commutative Smooth Surfaces

Michel Van den Bergh

November 2001 • Volume 154 • Number 734 (end of volume) • ISSN 0065-9266

American Mathematical Society
Providence, Rhode Island

2000 *Mathematics Subject Classification.*
Primary 16E40.

Library of Congress Cataloging-in-Publication Data

Bergh, M. Van den.
Blowing up of non-commutative smooth surfaces / Michel Van den Bergh.
p. cm. — (Memoirs of the American Mathematical Society, ISSN 0065-9266 ; no. 734)
"Volume 154, number 734 (end of volume)."
Includes bibliographical references and index.
ISBN 0-8218-2754-5 (alk. paper)
1. Noncommutative differential geometry. 2. Blowing up (Algebraic geometry) I. Title. II. Series.

QA3 .A57 no. 734
[QA641]
510 s—dc21
[516.3′6] 2001027925

Memoirs of the American Mathematical Society

This journal is devoted entirely to research in pure and applied mathematics.

Subscription information. The 2001 subscription begins with volume 149 and consists of six mailings, each containing one or more numbers. Subscription prices for 2001 are $494 list, $395 institutional member. A late charge of 10% of the subscription price will be imposed on orders received from nonmembers after January 1 of the subscription year. Subscribers outside the United States and India must pay a postage surcharge of $31; subscribers in India must pay a postage surcharge of $43. Expedited delivery to destinations in North America $35; elsewhere $130. Each number may be ordered separately; *please specify number* when ordering an individual number. For prices and titles of recently released numbers, see the New Publications sections of the *Notices of the American Mathematical Society.*

Back number information. For back issues see the *AMS Catalog of Publications.*

Subscriptions and orders should be addressed to the American Mathematical Society, P. O. Box 845904, Boston, MA 02284-5904. *All orders must be accompanied by payment.* Other correspondence should be addressed to Box 6248, Providence, RI 02940-6248.

Memoirs of the American Mathematical Society is published bimonthly (each volume consisting usually of more than one number) by the American Mathematical Society at 201 Charles Street, Providence, RI 02904-2294. Periodicals postage paid at Providence, RI. Postmaster: Send address changes to Memoirs, American Mathematical Society, P. O. Box 6248, Providence, RI 02940-6248.

This publication is indexed in *Science Citation Index*®, *SciSearch*®, *Research Alert*®, *CompuMath Citation Index*®, *Current Contents*®*/Physical, Chemical & Earth Sciences.*
Printed in the United States of America.

♾ The paper used in this book is acid-free and falls within the guidelines established to ensure permanence and durability.
Visit the AMS home page at URL: `http://www.ams.org/`

10 9 8 7 6 5 4 3 2 1 06 05 04 03 02 01

Contents

Chapter 1. Introduction 1
- 1.1. Motivation 1
- 1.2. Construction 3
- 1.3. General properties 5
- 1.4. Non-commutative Del-Pezzo surfaces 6
- 1.5. Exceptional simple objects 6
- 1.6. Non-commutative cubic surfaces 7
- 1.7. Acknowledgement 7

Chapter 2. Preliminaries on category theory 8

Chapter 3. Non-commutative geometry 10
- 3.1. Bimodules 10
- 3.2. Graded modules, bimodules and algebras 18
- 3.3. Quotients of the identity functor 19
- 3.4. Ideals in the identity functor 22
- 3.5. Quasi-schemes 26
- 3.6. Divisors 28
- 3.7. Proj 29
- 3.8. Condition "$\boldsymbol{\chi}$" and cohomological dimension 32
- 3.9. Higher inverse images 39
- 3.10. Algebras which are strongly graded modulo a Serre subcategory 40
- 3.11. The positive part of certain graded algebras. 41
- 3.12. Veronese subalgebras 42

Chapter 4. Pseudo-compact rings 44

Chapter 5. Cohen-Macaulay curves embedded in quasi-schemes 53
- 5.1. Preliminaries 53
- 5.2. Some computations 57
- 5.3. Completion of objects in $\mathbf{mod}(\boldsymbol{X})$ 60
- 5.4. Completion of bimodules 61
- 5.5. The category $\tilde{\mathcal{C}}_{f,p}$ 64
- 5.6. Completion of algebras 68
- 5.7. Multiplicities in the case that τ has infinite order 69

Chapter 6. Blowing up a point on a commutative divisor 72
- 6.1. Some ideals 72
- 6.2. Some Rees algebras 76
- 6.3. Definition of blowing up 77
- 6.4. The normal bundle 79

6.5. Birationality 80
6.6. The exceptional curve 82
6.7. The structure of the exceptional curve 84
6.8. The strict transform 88
6.9. A result on K_0 of some categories 91

Chapter 7. Derived categories 93
7.1. Generalities 93
7.2. Admissible compositions of morphisms between quasi-schemes 93

Chapter 8. The derived category of a non-commutative blowup 96
8.1. The formalism of semi-orthogonal decompositions 96
8.2. Generalities 97
8.3. Computation of some derived functors 98
8.4. The main theorem 101

Chapter 9. Some results on graded algebras and their sections 105
9.1. Generalities 105
9.2. The case of a blowing up 111

Chapter 10. Quantum plane geometry 113
10.1. Multiplicities of some objects 113
10.2. Classification of lines and conics 118

Chapter 11. Blowing up n points in an elliptic quantum plane 121
11.1. Derived categories 121
11.2. Exceptional simple objects 126

Chapter 12. Non-commutative cubic surfaces 129

Appendix A. Two-categories 132

Appendix B. Summary of notations 135

Appendix C. Index of terminology 138

Bibliography 139

Abstract

In this paper we will think of certain abelian categories with favorable properties as non-commutative surfaces. We show that under certain conditions a point on a non-commutative surface can be blown up. This yields a new non-commutative surface which is in a certain sense birational to the original one. This construction is analogous to blowing up a Poisson surface at a point of the zero-divisor of the Poisson bracket.

By blowing up ≤ 8 points in the elliptic quantum plane one obtains global non-commutative deformations of Del-Pezzo surfaces. For example blowing up six points yields a non-commutative cubic surface. Under a number of extra hypotheses we obtain a formula for the number of non-trivial simple objects on such non-commutative surfaces.

Received by the editor November 2, 1998, and in revised form July 10, 2000.
1991 *Mathematics Subject Classification.* Primary 16E40.
Key words and phrases. Blowing up, non-commutative geometry.
The author is a senior researcher at the FWO.
The author dedicates this monograph to Sarah and Bertold.

CHAPTER 1

Introduction

Throughout this paper k will be an algebraically closed field.

1.1. Motivation

Let X be a smooth connected surface over k and let q be a Poisson bracket on X. Since we are in the dimension two, q corresponds to a section of the anti-canonical bundle ω_X^*.

Let $p \in X$ and let $\alpha : \tilde{X} \to X$ be the blowup of X at p. From the fact that $\tilde{X}$ and X share the same function field it is easily seen that q extends to $\tilde{X}$ if and only if q vanishes at p. Denote the extended Poisson bracket by q' and let Y resp. T be the zero divisors of q and q'. One verifies that as divisors : $T = \alpha^{-1}(Y) - L$, where $L = \alpha^{-1}(p)$ is the exceptional curve. In particular T contains the strict transform $\tilde{Y}$ of Y, and if $p \in Y$ is simple then actually $T = \tilde{Y}$.

Our aim in this paper is to show that there exists a non-commutative version of this situation. That is we show that it is possible to view the blowup of a Poisson surface as the quasi-classical analogue of a blowup of a non-commutative surface. Our motivation for doing this is to provide a step in the ongoing project of classifying graded domains of low Gelfand-Kirillov dimension. Since the case of dimension two was completely solved in [**5**] the next interesting case will very likely be dimension three (leaving aside rings with fractional dimension which seem to be quite exotic). One may view three dimensional graded rings as homogeneous coordinate rings of non-commutative projective surfaces. Motivated by some heuristic evidence Mike Artin conjectures in [**4**] that, up to birational equivalence, there will be only a few classes, the largest one consisting of those algebras that are birational to a quantum $\mathbb{P}^2$ (see below).

Once a birational classification exists, one might hope that there would be some version of Zariski's theorem saying that if two (non-commutative) surfaces are birationally equivalent then they are related through a sequence of blowing ups and downs. With the current level of understanding it seems rather unlikely that Artin's conjecture or a non-commutative version of Zariski's theorem will be proved soon, but this paper provides at least one piece of the puzzle.

This being said, it is perhaps the right moment to point out that in this paper we won't really define the notion of a non-commutative surface. Instead we first introduce non-commutative schemes (or quasi-schemes, to follow the terminology of [**28**]). These will simply be abelian categories having sufficiently nice homological properties. Then we will impose a few convenient additional hypotheses which would hold for a commutative smooth surface (see §5.1).

To fix ideas we will first discuss two particular cases of quasi-schemes. If R is a ring then $\operatorname{Spec} R$ is the category of right R-modules (the "affine case"). If $A = A_0 \oplus A_1 \oplus \cdots$ is a graded ring then $\operatorname{Proj} A$ is (roughly) the category of graded right A-modules, modulo the modules with right bounded grading (the "projective case").

Let us first consider the affine case. Assume that R is a finitely generated k-algebra and let C be the commutator ideal. C is the natural analog of the zero divisor of a Poisson bracket. Now $\operatorname{Spec} R/C$ is a commutative affine scheme and a k-point in $\operatorname{Spec} R/C$ corresponds to a maximal ideal m in R with $R/m = k$. Hence a natural idea is to define the blowup of $\operatorname{Spec} R$ at p as $\operatorname{Proj} D$ where D is the Rees algebra associated to m,

$$D = R \oplus m \oplus m^2 \oplus m^3 \oplus \cdots$$

It is easily seen however that this definition is faulty. Consider the following example [**4**] $R = k\langle x, y\rangle/(yx - xy - y)$, $m = (x, y)$. Then $m^n = (x^n, y)$. Hence the analog of the exceptional curve

$$D/mD = R/m \oplus m/m^2 \oplus m^2/m^3 \oplus$$

is isomorphic to $k[x]$. Thus $\operatorname{Proj} D/m$ is a point, whereas intuitively we would expect it to be one-dimensional in some sense.

It turns out however that in this example one can use a certain twisting of the Rees algebra which yields a reasonably behaved blowing up. This is based on the observation that the commutator ideal $C = (y)$ is an invertible R-bimodule. Let J be its inverse and put $I = mJ$. Then we define I^n as the image of $I^{\otimes n}$ in $J^{\otimes n}$ and we define the modified Rees algebra D as

$$D = R \oplus I \oplus I^2 \oplus I^3 \oplus \cdots$$

The blowup of $\operatorname{Spec} R$ at p is now defined as $\operatorname{Proj} D$ for this new D. We refer the reader to [**4**] for a detailed workout of this example. However we will indicate how one finds the analog of the exceptional curve. Let τ be the automorphism of R given by $a \mapsto y^{-1}ay$. For M an R-bimodule let M_τ be the bimodule whose right R-action is twisted by *tau* (i.e. $m \cdot r = m\tau(r)$). Then $J = R_\tau$ and hence $I^n = m\tau(m)\cdots\tau^{n-1}(m)_{\tau^n}$. Put $L = D/\tau^{-1}(m)D$. One now verifies that $\dim L_u = u + 1$. So L plays the role of the exceptional curve. Note however that L is a right D-module but *not* a left module. In retrospect this was to be expected since, as we have said in the first paragraph, if we blow up a Poisson surface, then the extended Poisson bracket (if it exists) will in general not vanish on the exceptional curve.

This example indicates the way to go for rings whose commutator ideal is invertible. The latter hypothesis is not unreasonable since if we look at the case of a Poisson surface then we see that we expect a non-commutative smooth surface to contain a commutative curve. Additional motivation comes from considering the local rings $k\langle\langle x, y\rangle\rangle/(\phi)$ with ϕ a (non-commutative) formal power series whose lowest degree term is a non-degenerate quadratic tensor in x, y. These local rings are the non-commutative analogs of complete two dimensional regular local rings and one verifies that their commutator ideal is indeed invertible (see e.g. [**36**]).

There is one important hitch however. The commutator ideal is not invariant under Morita equivalence! This indicates that it is important to develop the theory in a more category-theoretical frame work. This will make it possible to talk about non-commutative schemes containing a commutative curve, without refering to rings or ideals at all.

To stress this point even more let us consider the case of graded rings. In [**7**] Artin and Schelter introduced so-called regular rings. These are basically graded rings which have the Hilbert series of three dimensional polynomial rings, together with a few other reasonable properties. They were classified in [**7, 8, 9**] and also, with different methods, in [**13**]. Let A be such a regular ring. We view $X = \operatorname{Proj} A$ as a quantum $\mathbb{P}^2$. Since on $\mathbb{P}^2$ the anti-canonical sheaf has degree three, the zero divisor of a Poisson bracket will be a (possibly singular and non-reduced) elliptic curve. Therefore we would also expect $X = \operatorname{Proj} A$ to contain an elliptic curve in some reasonable sense. It turns out that this is indeed true! It was shown in [**7, 10, 8**] that A contains a normal element g in degree three such that $\operatorname{Proj} A/gA$ is equivalent to the category of quasi-coherent sheaves over an elliptic curve Y. Thus if we actually identify Y with $\operatorname{Proj} A/gA$ then $Y \hookrightarrow X$.

Now let $p \in Y$. The previous discussion suggests that it should be possible to blow up p. However it is not clear how to proceed. Under the inclusion $Y \hookrightarrow X$, p corresponds to a so-called point module [**9**] over A. This is by definition a graded right module, which is generated in degree zero and which is one-dimensional in every degree. However such a point module is only a right module and hence it cannot be used to construct a Rees algebra.

Our solution is to construct the Rees algebra directly over $\operatorname{Proj} A$. To do this we have to invoke the theory of monads [**24**]. Since we only consider monads satisfying a lot of additional hypotheses we prefer to call our monads algebras. This is at variance with the use of "algebra" in the theory of categories [**24**] but in our context it seems reasonable. For us an algebra over an abelian category $\mathcal{C}$ is in principle an algebra object in the monoidal category of right exact functors from $\mathcal{C}$ to itself. There are however some technical problems with this so we end up using a less intuitive definition (see below).

The importance of monads in non-commutative algebraic geometry was noticed by various people, in particular by Rosenberg. See for example [**28, 23**]. In the last chapter of his book Rosenberg actually defines a blow up of an arbitrary "closed" subcategory of an abelian category. While this definition is also in terms of monads, it is as far as I can see, somewhat different from ours. To see this let us again consider the affine case. Then Rosenberg's construction is in terms of the functor $M \mapsto Mm$, which is not right exact. If we replace this functor by $M \mapsto M \otimes_R m$ then one would get the Proj of the ordinary Rees algebra of R, which (depending on what one wants to achieve) might not be the right answer (as we have shown above).

1.2. Construction

Following [**28**] we introduce the notion of a quasi-scheme. For us this will be a Grothendieck category (that is : an abelian category with a generator and exact direct limits). However we tend to think of quasi-schemes as geometric objects, so we denote them by roman capitals $X, Y, \ldots$. If we really refer to the category represented by a quasi-scheme X then we write $\operatorname{Mod}(X)$. Note that in fact $X = \operatorname{Mod}(X)$, but it is very useful to nevertheless make this notational distinction since it allows us to introduce other notations in a consistent way. For example we will denote the noetherian objects in $\operatorname{Mod}(X)$ by $\operatorname{mod}(X)$. Furthermore we can absorb additional structure into the symbol X (such as a morphism to a base quasi-scheme) which is not related to $\operatorname{Mod}(X)$. This would be awkward without the two different notations X and $\operatorname{Mod}(X)$.

A morphism $\alpha : X \to Y$ between quasi-schemes will be a right exact functor $\alpha^* : \mathrm{Mod}(Y) \to \mathrm{Mod}(X)$ possessing a right adjoint (denoted by α_*). In this way the quasi-schemes form a category (more precisely a two-category, see Appendix A).

If X is a quasi-compact quasi-separated commutative scheme then $\mathrm{Mod}(X)$ will be the category of quasi-coherent sheaves on X. It is proved in [**32**] that this is a Grothendieck category. Rosenberg in [**29**] has proved a reconstruction theorem which allows one to recover X from $\mathrm{Mod}(X)$ (generalizing work of Gabriel in the noetherian case). He has also announced that the functor which assigns to a commutative scheme its associated quasi-scheme is fully faithful if we work over $\mathrm{Spec}\,\mathbb{Z}$.

Let X be a quasi-scheme. We think of objects in $\mathrm{Mod}(X)$ as sheaves of right modules on X. However to define algebras on X, it is clear that we need bimodules on X (see [**34**] for the case where X is commutative). Let us for the moment define a bimodule on X as a right exact functor from $\mathrm{Mod}(X)$ to itself commuting with direct limits. Then the category of bimodules is monoidal (the tensor product being given by composition) and hence we can define algebra objects. Let $\mathcal{A}$ be such an algebra object. It is routine to define an abelian category $\mathrm{Mod}(\mathcal{A})$ of $\mathcal{A}$-modules. So this seems like a reasonable starting point for the theory.

However a difficulty emerges if one wants to define Rees algebras. As we have seen, the main point is to take the sum of the I^n for some subbimodule I of an invertible bimodule $\mathcal{L}$. I^n was defined as the image of $I^{\otimes n} \to \mathcal{L}^{\otimes n}$. Unfortunately to take an image one needs an abelian category, and I don't see how to prove that the above definition of a bimodule yields an abelian category, even if we drop the requirement that bimodules should commute with direct limits. In this paper we sidestep this difficulty by defining the category of bimodules on X as the opposite category of the category of left exact functors from $\mathrm{Mod}(X)$ to itself. Since left exact functors are determined by their values on injectives, they trivially form an abelian category. In this way one can define Rees algebras in reasonable generality (see Definition 3.4.13).

We will say that a quasi-scheme X is noetherian if $\mathrm{Mod}(X)$ is locally noetherian. That is, if $\mathrm{Mod}(X)$ is generated by $\mathrm{mod}(X)$. As already has been indicated above, in this paper we will study a noetherian quasi-scheme X which contains a commutative curve Y as a divisor. To make this more precise we denote the identity functor on $\mathrm{Mod}(X)$ by o_X. This is an algebra on X such that $\mathrm{Mod}(o_X) = \mathrm{Mod}(X)$. We will assume that o_X contains an invertible subbimodule $o_X(-Y)$ such that $\mathrm{Mod}(o_X/o_X(-Y))$ is equivalent to $\mathrm{Mod}(Y)$.

We also need some sort of smoothness condition on X. Since it is obviously sufficient to impose this in a neighborhood of Y, we assume that every object in $\mathrm{Mod}(Y)$ has finite injective dimension in $\mathrm{Mod}(X)$. This is the same setting as in [**36**], albeit cast in a somewhat different language.

Now $p \in Y$ defines a subbimodule m_p of o_X which is the analog of the maximal ideal corresponding to p. We put $I = m_p o_X(Y)$. Define

$$\mathcal{D} = o_X \oplus I \oplus I^2 \oplus \cdots$$

This is the Rees algebra associated to I. We define the blowup $\tilde{X}$ of X at p as $\mathrm{Proj}\,\mathcal{D}$.

Two remarks are in order here.

1. It is easy to see that the above definition gives the correct result in the commutative case.
2. The definition of non-commutative blowup seems to depend upon the curve Y. One can show however that the result is independent of the choice of Y. In fact one can even give a reformulation of the definition which does not involve Y. This will be explained in [**33**]. To prove that the blowup has the correct properties one still seems to need the curve Y.

1.3. General properties

A large part of this paper is devoted to proving that $\tilde{X}$ satisfies similar properties as X and furthermore that we have obtained an analogue of the blowup of a commutative surface. However before we start we need to have a better understanding of the formal neighborhood of a point $p \in Y$. This was in fact already done in [**36**]. The answer is given in terms of certain topological rings (see Theorem 5.1.4 for precise results). It turns out that the formal local structure of p depends heavily on a certain automorphism τ of Y which was also introduced in [**36**]. To be more precise we define the normal bundle of Y in X as $o_X(Y)/o_X$. This bimodule defines an autoequivalence of $\operatorname{Mod}(Y)$ and by a result in [**6**] such an autoequivalence must necessarily be of the form $\tau_*(- \otimes_{\mathcal{O}_Y} \mathcal{N})$ where $\mathcal{N}$ is a line bundle on Y and τ is an automorphism of Y. In particular if τ has infinite order (and hence $p \in Y$ is smooth) then the formal local structure of p is given by the ring of doubly infinite lower triangular matrices over the ring $\hat{\mathcal{O}}_{Y,p}$. In particular this is independent of p and Y (as is the case for the completion at a smooth point on a commutative scheme).

In this paper we complete the results in [**36**] by showing that various completion functors are exact. For this we refer the reader to §5. An interesting application of completion is given in section §5.7. In this section we define (roughly) the multiplicity at p and the points infinitely near to p of an object in $\operatorname{mod}(X)$ in the case that the τ-orbit of p is infinite.

As a starting point for the study of $\tilde{X}$ we construct a commutative diagram of quasi-schemes.

$$\begin{array}{ccc} \tilde{Y} & \xrightarrow{\beta} & Y \\ {\scriptstyle i}\downarrow & & \downarrow{\scriptstyle j} \\ \tilde{X} & \xrightarrow[\alpha]{} & X \end{array}$$

where the vertical arrows are inclusions. $\tilde{X}$ is again a noetherian quasi-scheme (Theorem 6.3.1). $\tilde{Y}$ is a commutative curve which plays the role of the strict transform of Y. $\tilde{Y}$ is again a divisor in X and every object in $\operatorname{Mod}(\tilde{Y})$ has finite injective dimension in $\operatorname{Mod}(\tilde{X})$ (see Theorem 6.6.3).

Associated to the point $p \in Y$ there is a simple object $\mathcal{O}_p$ in $\operatorname{Mod}(Y) \subset \operatorname{Mod}(X)$. We define $\mathcal{O}_L$ as $\alpha^*(\mathcal{O}_p)$ and we consider $\mathcal{O}_L$ as the structure sheaf on the exceptional curve in $\tilde{X}$. In fact following a recipe given in [**30**] we can define a category $\operatorname{Mod}(L)$. Roughly $\operatorname{Mod}(L)$ is generated by subquotients of direct sums of twists of $\mathcal{O}_L$. In this way we can speak of the quasi-scheme L. It follows from Proposition 6.5.2 (together with Corollary 6.7.4) that if we view $\operatorname{Mod}(X)$ modulo the objects supported on the τ-orbit of p, and $\operatorname{Mod}(\tilde{X})$ modulo the objects supported on L then we obtain equivalent categories. This is the obvious analogue of the situation in the commutive case where $X - p$ and $\tilde{X} - L$ are isomorphic (and hence in particular X and $\tilde{X}$ are birational).

In section §6.7 we give a precise description of $\mathrm{Mod}(L)$ (using results of [**30**] in the case that τ has infinite order). It will follow that $\mathrm{Mod}(L)$ is very closely related to the category of quasi-coherent sheaves on $\mathbb{P}^1$, illustrating again the analogy with the commutative case.

In Chapter §7 we compute the derived category of $\tilde{X}$. Our main result is that this derived category has a semi-orthogonal decomposition given by the derived category of X and the derived category of k. This generalizes a result by Orlov [**27**].

1.4. Non-commutative Del-Pezzo surfaces

The results starting from Chapter 9 are inspired by the construction in the commutative case of (most) Del-Pezzo surfaces by blowing up a collection of points in $\mathbb{P}^2$. Let $(Y, \sigma, \mathcal{L})$ be an elliptic triple $Y \subset \mathbb{P}^2$ as in [**8**]. We assume that Y is smooth and that σ is a translation of infinite order. Let A be the regular algebra associated to this triple [**8**] and let $X_1 = \operatorname{Proj} A$. As above we consider X_1 as a quantum version of $\mathbb{P}^2$. The curve Y is contained as a divisor in X_1 and τ is equal in this case to σ^3. We choose points $p_1, \dots, p_n \in Y$ $(n \leq 8)$ and we construct a commutative diagram of quasi-schemes

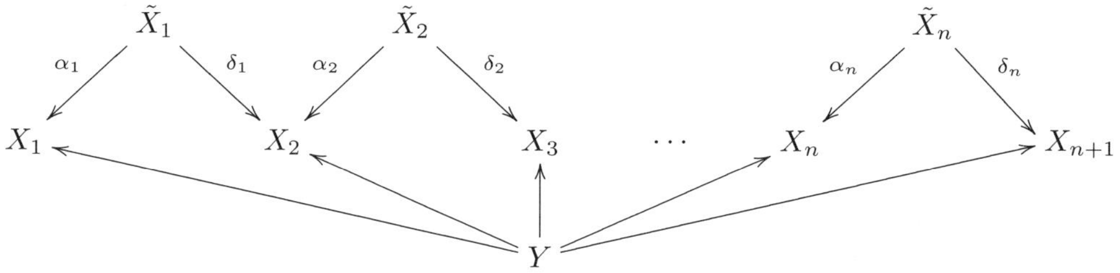

Here the morphism α_i is the blowup of X_i at p_i. Morally X_{i+1} is constructed from $\tilde{X}_i$ by putting $X_{i+1} = \operatorname{Proj}\left(\bigoplus_n H^0(X_i, o_{X_i}(nY))\right)$ (actually for simplicity X_{i+1} is constructed using a slightly different method (see §9.2), which is easily seen to be equivalent). The point is that in the commutative case δ_i would be an isomorphism if the points $p_1, \dots, p_n$ are in general position (this follows from the Nakai criterion for ampleness, see [**19**, Thm V.1.10, p365]). In the non-commutative case δ_i will not in general be an isomorphism. However we show in Theorem 11.1.3 that δ_i yields a derived equivalence, under suitable general position hypotheses.

1.5. Exceptional simple objects

One of the aims of these notes is to classify (or rather count) the simple objects in $\mathrm{Mod}(X_i)$ which are not of the form $\mathcal{O}_q$ for some $q \in Y$. We call such simple objects "exceptional" because, firstly, they do not always exist, and secondly when they exist they are not easy to construct or to count.

Using some results on geometry in the projective quantum plane (§10) together with the above results on derived categories we obtain in Theorem 11.2.1 a formula for the number of exceptional simple objects in $\mathrm{Mod}(X_{i+1})$ (if $n \leq 6$, τ has infinite order and suitable general position hypotheses hold). It turns out that the number of exceptional simple objects depends in a *very* sensitive way upon the position of the points $p_1, \dots, p_n$. For example if $n = 6$ (and the other hypotheses are satisfied) then our formula yields that there may be between 0 and 6 exceptional simple objects in $\mathrm{Mod}(X_7)$.

1.6. Non-commutative cubic surfaces

Being near the end of this introduction we now indicate our original motivation for starting this project. It concerns a problem which is not quite completely solved but which at least has become more tractable.

As above let $(Y, \sigma, \mathcal{L})$ be an elliptic triple with $Y \in \mathbb{P}^2$ and let A be the associated three dimensional Artin-Schelter regular algebra. It is easy to show that the representation theory of A is fairly trivial. At a certain point Lieven Le Bruyn (see for example [**22**]) suggested that one could obtain more interesting representation theories by considering filtered rings C such that $\mathrm{gr}\, C = A$. This was motivated by the example of enveloping algebras of Lie algebras which are related in a similar way to polynomial algebras.

Instead of studying the filtered rings C, it is easier to study their Rees rings $D = \oplus_n C_n$. These are characterized as the $\mathbb{N}$-graded rings containing a regular central element t in degree one such that $D/tD = A$. Such graded rings were studied in [**21**] but not much progress was made towards their representation theory.

As usual a first step in the study of the representation theory of D is the construction of a "Casimir" element. Indeed A contains a canonical central element g in degree 3 and by a computer computation one can show that g lifts to a central element G in D. Then instead of studying D one may study the central quotients $D_\mu = D/(G + \mu t^3)$, $\mu \in k$. The homogeneous coordinate rings of the D_μ may be considered as quantum versions of cubic surfaces in $\mathbb{P}^3$.

A well known result in commutative algebraic geometry states that a cubic surface in $\mathbb{P}^3$ is obtained by blowing up six points in $\mathbb{P}^2$, so one may ask whether this is also true in the noncommutative case. I have not been able to show this in general but a converse result is obtained in Chapter 12. We show that if we blow up six points in the elliptic quantum plane then the resulting quasi-scheme is contained as a cubic divisor in a quantum $\mathbb{P}^3$.

Very recently Mike Artin has explained to me that one can probably obtain the complete analogue of the commutative result by deformation theory. Indeed $Z = \mathrm{Proj}\, D_\mu$ can be obtained as a deformation of a cubic surface Z_0 in $\mathbb{P}^3$. In Z_0 we can choose six mutually skew exceptional curves [**25**, Thm 24.2, p119] and since the structure sheaves of these exceptional curves have no higher Ext's they deform well. So we should find on Z six corresponding exceptional curves. Then we can blow these down, for example using the procedure exhibited in [**35**]. To carry out this program there are quite a few technical details that remain to be filled in. There are some recent notes by Mike Artin on specializing birational equivalences in the non-commutative case.

1.7. Acknowledgement

The author wishes to thank Mike Artin and Paul Smith for stimulating discussions about the material in earlier versions of this manuscript and about noncommutative geometry in general. I also wish to thank Paul Smith for showing me the interesting preprint [**30**] which contains related results.

CHAPTER 2

Preliminaries on category theory

Our main references for categories are [**16, 17, 24, 31**]. If nothing is specified then categories will have small homsets. This will not always be the case for some secondary categories such as categories of bimodules (see below). However limits and colimits are always taken over small sets. In particular complete and cocomplete will refer to the existence of small limits and colimits. When we speak of a direct limit, we mean a colimit over a *directed* set. A similar convention applies to inverse limits.

We will often implicitly use the fact that in an abelian category

$$\oplus_{i\in I} Y_i = \varinjlim_{\substack{j\in J\\ J\subset I \text{ finite}}} Y_j \qquad \prod_{i\in I} Y_i = \varprojlim_{\substack{j\in J\\ J\subset I \text{ finite}}} Y_j \tag{2.1}$$

Thus an additive functor commuting with direct limits commutes with coproducts. A similar statement applies to products.

We use the following specialized version of the standard adjoint functor theorems [**24**, Corollary IV.8, p126].

THEOREM 2.1. *Let $\mathcal{C}$, $\mathcal{D}$ be abelian categories and let $F : \mathcal{D} \to \mathcal{C}$, $G : \mathcal{C} \to \mathcal{D}$ be respectively a right and a left exact functor. Then*

1. *If $\mathcal{C}$ is complete and has a cogenerator then G has a left adjoint if and only if G commutes with products.*
2. *(Dual version) If $\mathcal{D}$ is cocomplete and has a generator then F has a right adjoint if and only if F commutes with coproducts.*

Most of the abelian categories we use will be *Grothendieck categories*. These are abelian categories which have a generator and exact direct limits. We use the following results which are well-known [**17, 31**].

THEOREM 2.2. *Let $\mathcal{C}$ be a Grothendieck category. Then*

1. *$\mathcal{C}$ has products (not necessarily exact).*
2. *$\mathcal{C}$ has enough injectives.*
3. *The product of the injective hulls of the quotients of a fixed generator is an injective cogenerator.*

A Grothendieck category which is generated by noetherian objects is called *locally noetherian*. Such categories were studied by Gabriel in [**16**]. One important property they have is the following.

THEOREM 2.3. *In a locally noetherian category the direct limit of injective objects is injective.*

A *noetherian* category is an abelian category in which every object is noetherian. One has [**16**, III, Th 1]

THEOREM 2.4. *Every noetherian category $\mathcal{C}$ is equivalent to the category of noetherian objects in a locally noetherian category $\tilde{\mathcal{C}}$. $\tilde{\mathcal{C}}$ is characterized up to equivalence by this property.*

A *Serre subcategory* $\mathcal{S}$ of an abelian category $\mathcal{C}$ is by definition a full subcategory which is closed under subquotients and extensions. In that case there exists a quotient category $\mathcal{C}/\mathcal{S}$ (as is well-known there is a small set theoretic difficulty in the sense that the Hom-"sets" in $\mathcal{C}/\mathcal{S}$ might not be sets). If $\mathcal{C}$ is a Grothendieck category then we say that $\mathcal{S}$ is localizing if $\mathcal{S}$ is a Serre subcategory which is closed under direct limits. In that case $\mathcal{C}/\mathcal{S}$ is also a Grothendieck category (and in particular the above mentioned set theoretic difficulty goes away). Furthermore the quotient functor $\pi : \mathcal{C} \to \mathcal{C}/\mathcal{S}$ has a right adjoint which we will usually denote by ω. The composition $\omega\pi$ is called the *localization functor* and will be denoted by $\tilde{(-)}$.

In this paper we basically work with the category of Grothendieck categories, the morphisms being certain functors. This is an example of a two-category. Although we usually only implicitly use this concept, we explain some basic notions in Appendix A.

CHAPTER 3

Non-commutative geometry

In this paper we will sometimes work with categories which do not come from categories of modules over (graded) rings. Therefore in this chapter we introduce some rudiments of a formalism which may be used to imitate some of the more elementary features of commutative algebraic geometry. This chapter is closely related to [**28**][**34**].

3.1. Bimodules

Below $\mathcal{C}$, $\mathcal{D}$, $\mathcal{E}$, ... will be abelian categories. We define the following categories.

$$\begin{aligned} \mathcal{L}(\mathcal{D},\mathcal{C}) &= \{\text{left exact functors } \mathcal{D} \to \mathcal{C}\} \\ \mathrm{BIMOD}(\mathcal{C}-\mathcal{D}) &= \mathcal{L}(\mathcal{D},\mathcal{C})^{\mathrm{opp}} \\ \mathrm{Bimod}(\mathcal{C}-\mathcal{D}) &= \{\mathcal{M} \in \mathrm{BIMOD}(\mathcal{C}-\mathcal{D}) \mid \mathcal{M} \text{ has a left adjoint }\} \end{aligned}$$

Objects in $\mathrm{BIMOD}(\mathcal{C}-\mathcal{D})$ will be called *weak* $\mathcal{C}$-$\mathcal{D}$ *bimodules*, whereas objects in $\mathrm{Bimod}(\mathcal{C}-\mathcal{D})$ will simply be called $\mathcal{C}$-$\mathcal{D}$ *bimodules*.

PROPOSITION 3.1.1. 1. $\mathrm{BIMOD}(\mathcal{C}-\mathcal{D})$, $\mathrm{Bimod}(\mathcal{C}-\mathcal{D})$ *have finite colimits, and the inclusion* $\mathrm{Bimod}(\mathcal{C}-\mathcal{D}) \to \mathrm{BIMOD}(\mathcal{C}-\mathcal{D})$ *preserves those colimits.*

2. *If* $\mathcal{C}$ *is complete then* $\mathrm{BIMOD}(\mathcal{C}-\mathcal{D})$ *is cocomplete. If* $\mathcal{D}$ *is in addition cocomplete then so is* $\mathrm{Bimod}(\mathcal{C}-\mathcal{D})$ *and furthermore the inclusion* $\mathrm{Bimod}(\mathcal{C}-\mathcal{D}) \to \mathrm{BIMOD}(\mathcal{C}-\mathcal{D})$ *preserves colimits.*
3. *If* $\mathcal{D}$ *has enough injectives then* $\mathrm{BIMOD}(\mathcal{C}-\mathcal{D})$ *is an abelian category.*
4. *If* $\mathcal{D}$ *is complete and has a cogenerator then an object in* $\mathrm{BIMOD}(\mathcal{C}-\mathcal{D})$ *is in* $\mathrm{Bimod}(\mathcal{C}-\mathcal{D})$ *iff it commutes with products.*
5. *If* $\mathcal{D}$ *is complete and has an injective cogenerator then an object in the category* $\mathrm{BIMOD}(\mathcal{C}-\mathcal{D})$ *is in* $\mathrm{Bimod}(\mathcal{C}-\mathcal{D})$ *iff it commutes with products when evaluated on injectives.*
6. *If* $\mathcal{D}$ *is complete and has an injective cogenerator and products are exact in* $\mathcal{C}$ *then* $\mathrm{Bimod}(\mathcal{C}-\mathcal{D})$ *is an abelian subcategory of* $\mathrm{BIMOD}(\mathcal{C}-\mathcal{D})$.

PROOF. 1. To show that $\mathrm{BIMOD}(\mathcal{C}-\mathcal{D})$ has finite colimits we have to show that the opposite category $\mathcal{L}(\mathcal{D},\mathcal{C})$ has finite limits. Now one easily verifies that for a functor $G : I \to \mathcal{L}(\mathcal{D},\mathcal{C})$ with I a finite category

$$G'(M) = \varprojlim_i G(i)(M)$$

defines the inverse limit of G in $\mathcal{L}(\mathcal{D},\mathcal{C})$.

Assume now that the $G(i)$ are in $\mathrm{Bimod}(\mathcal{C}-\mathcal{D})$ so they have left adjoints $F(i)$. Define F' by

$$F'(N) = \varinjlim_i F(i)(N)$$

One easily verifies that F' is a left adjoint to G' and hence $G' \in \mathrm{Bimod}(\mathcal{C} - \mathcal{D})$.
2. This is similar to (1).
3. Let $\mathrm{Inj}(\mathcal{D})$ denote the additive category of injectives in $\mathcal{D}$. Then $\mathcal{L}(\mathcal{D}, \mathcal{C})$ is equivalent to $\mathrm{Funct}(\mathrm{Inj}(\mathcal{D}), \mathcal{C})$, where "Funct" denotes the category of additive functors. It is now clear that $\mathcal{L}(\mathcal{D}, \mathcal{C})$ inherits the property of being an abelian category from $\mathcal{C}$.
4. This follows from Theorem 2.1.
5. This follows from (4), using the fact that products are left exact.
6. We have to show that $\mathrm{Bimod}(\mathcal{C} - \mathcal{D})$ is closed under kernels. Equivalently the subcategory of $\mathrm{Funct}(\mathrm{Inj}(\mathcal{D}), \mathcal{C})$ of functors commuting with products has to be closed under cokernels.

   Let $G_1 \to G_2$ be a morphism of functors in $\mathrm{Funct}(\mathcal{D}, \mathcal{C})$ commuting with products and let $G_3 = \mathrm{coker}(G_1 \to G_2)$. Then for $E \in \mathrm{Inj}(\mathcal{D})$ we have $G_3(E) = \mathrm{coker}(G_1(E) \to G_2(E))$ and using the fact that products are exact one easily obtains that G_3 commutes with products. □

We will now introduce some more suggestive notations for dealing with $\mathrm{BIMOD}(\mathcal{C} - \mathcal{D})$.

If $\mathcal{M} \in \mathrm{BIMOD}(\mathcal{C} - \mathcal{D})$ then we denote the corresponding functor in $\mathcal{L}(\mathcal{D}, \mathcal{C})$ by $\mathcal{H}om_{\mathcal{D}}(\mathcal{M}, -)$. More or less by definition we have the following facts.

PROPOSITION 3.1.2. 1. *$\mathcal{H}om_{\mathcal{D}}(-,-)$ is left exact in its two arguments.*
2. *If $\mathcal{C}$ is complete then $\mathcal{H}om_{\mathcal{D}}(-,-)$ transforms colimits in its first argument into limits.*
3. *If $\mathcal{M} \in \mathrm{Bimod}(\mathcal{C} - \mathcal{D})$ then $\mathcal{H}om_{\mathcal{D}}(\mathcal{M}, -)$ transforms limits in its second argument into limits.*
4. *If $\mathcal{D}$ has enough injectives and $E \in \mathrm{Inj}(\mathcal{D})$ then $\mathcal{H}om_{\mathcal{D}}(-, E)$ is exact.*

PROOF. 1. That $\mathcal{H}om_{\mathcal{D}}(-,-)$ is left exact in its second argument is by definition. That it is left exact in its first argument follows from the explicit construction of colimits (and hence of cokernels) in the proof of Prop. 3.1.1(1).
2. This follows as in the proof of Prop. 3.1.1(2).
3. If $\mathcal{M} \in \mathrm{Bimod}(\mathcal{C} - \mathcal{D})$ then $\mathcal{H}om_{\mathcal{D}}(\mathcal{M}, -)$ has by definition a left adjoint. Hence it commutes with limits.
4. This follows from the explicit structure of an abelian category on $\mathrm{BIMOD}(\mathcal{C} - \mathcal{D})$ given by the proof Prop. 3.1.1(3). □

We write composition of functors

$$\mathrm{BIMOD}(\mathcal{C} - \mathcal{D}) \times \mathrm{BIMOD}(\mathcal{D} - \mathcal{E}) \to \mathrm{BIMOD}(\mathcal{C} - \mathcal{E})$$

as $- \otimes_{\mathcal{D}} -$. In this way we obtain for $\mathcal{M} \in \mathrm{BIMOD}(\mathcal{C} - \mathcal{D})$, $\mathcal{N} \in \mathrm{BIMOD}(\mathcal{D} - \mathcal{E})$ the satisfying formula

$$\mathcal{H}om_{\mathcal{E}}(\mathcal{M} \otimes_{\mathcal{D}} \mathcal{N}, -) = \mathcal{H}om_{\mathcal{D}}(\mathcal{M}, \mathcal{H}om_{\mathcal{E}}(\mathcal{N}, -))$$

Again more or less by definition we obtain the following properties.

PROPOSITION 3.1.3. 1. *$- \otimes_{\mathcal{D}} -$ is right exact in its two arguments.*
2. *If $\mathcal{C}$ is complete then $- \otimes_{\mathcal{D}} -$ preserves colimits in its first argument.*
3. *If $\mathcal{D}$ is complete and if $\mathcal{M} \in \mathrm{Bimod}(\mathcal{C} - \mathcal{D})$ then $\mathcal{M} \otimes_{\mathcal{D}} -$ preserves colimits in its second argument.*

4. $-\otimes_{\mathcal{D}}-$ *sends* $\text{Bimod}(\mathcal{C}-\mathcal{D})\times\text{Bimod}(\mathcal{D}-\mathcal{E})$ *to* $\text{Bimod}(\mathcal{C}-\mathcal{E})$.

If $F:\mathcal{E}\to\mathcal{F}$ is fully faithful then we define the *essential image* of F as the full subcategory of $\mathcal{F}$ consisting of objects isomorphic to objects of the form FE with $E\in\mathcal{E}$.

Now define

$$\text{MOD}(\mathcal{C})=\text{BIMOD}(\mathbf{Ab}-\mathcal{C})$$

The functor $M\mapsto \text{Hom}_{\mathcal{C}}(M,-)$ defines a full faithful embedding of $\mathcal{C}$ in $\text{MOD}(\mathcal{C})$. Throughout we will identify $\mathcal{C}$ with its essential image under this embedding.

We now obtain the following alternative "characterization" of $\text{Bimod}(\mathcal{C}-\mathcal{D})$ inside $\text{BIMOD}(\mathcal{C}-\mathcal{D})$.

PROPOSITION 3.1.4. *Let* $\mathcal{M}\in\text{BIMOD}(\mathcal{C}-\mathcal{D})$. *Then* $\mathcal{M}$ *is in* $\text{Bimod}(\mathcal{C}-\mathcal{D})$ *if and only if the functor*

$$-\otimes_{\mathcal{C}}\mathcal{M}:\text{MOD}(\mathcal{C})\to\text{MOD}(\mathcal{D})$$

sends $\mathcal{C}$ *to* $\mathcal{D}$.

A few concepts from the theory of bimodules over rings can be generalized to our setting. We denote the derived functors of $\mathcal{H}om_{\mathcal{D}}(-,-)$ in the second argument by $\mathcal{E}xt^i_{\mathcal{D}}(-,-)$ (if they exist). Assume that $\mathcal{E}$ has enough injectives. We say that $\mathcal{N}\in\text{BIMOD}(\mathcal{D}-\mathcal{E})$ is *(left) flat* if $\mathcal{H}om_{\mathcal{E}}(\mathcal{N},-)$ preserves injectives. More generally for $\mathcal{M}\in\text{BIMOD}(\mathcal{C}-\mathcal{D})$ and $\mathcal{N}\in\text{BIMOD}(\mathcal{D}-\mathcal{E})$ we define $\mathcal{T}or_i^{\mathcal{D}}(\mathcal{M},\mathcal{N})$ as the object in $\text{BIMOD}(\mathcal{C}-\mathcal{D})$ satisfying

$$\mathcal{H}om_{\mathcal{E}}(\mathcal{T}or_i^{\mathcal{D}}(\mathcal{M},\mathcal{N}),E)=\mathcal{E}xt^i_{\mathcal{D}}(\mathcal{M},\mathcal{H}om_{\mathcal{E}}(\mathcal{N},E)) \tag{3.1}$$

for every injective E of $\mathcal{E}$. A similar definition holds for $\mathcal{M}\in\mathcal{D}$.

PROPOSITION 3.1.5. *Assume that* $\mathcal{E}$ *has enough injectives.*

1. $\mathcal{T}or_i^{\mathcal{D}}(-,-)$ *is a* δ*-functor in its two arguments.*
2. *An object* $\mathcal{N}$ *in* $\text{BIMOD}(\mathcal{D}-\mathcal{E})$ *is flat if and only if* $\mathcal{T}or_1^{\mathcal{D}}(\mathcal{M},\mathcal{N})=0$ *for all* $\mathcal{M}\in\mathcal{D}$.
3. *If* $\mathcal{N}$ *in* $\text{BIMOD}(\mathcal{D}-\mathcal{E})$ *is flat then* $\mathcal{T}or_i^{\mathcal{D}}(\mathcal{M},\mathcal{N})=0$ *for every object* $\mathcal{M}$ *in* $\text{BIMOD}(\mathcal{C}-\mathcal{D})$.

PROOF. All statements follow directly from the definitions. □

Note that (1) implies that if $\mathcal{T}or_1^{\mathcal{D}}(-,\mathcal{N})=0$ then $-\otimes_{\mathcal{D}}\mathcal{N}$ is exact. We don't know if the converse of this statement holds.

Assume that $\mathcal{C}$, $\mathcal{D}$ have colimits. Then we will say that $\mathcal{M}\in\text{BIMOD}(\mathcal{C}-\mathcal{D})$ is coherent if $\mathcal{H}om_{\mathcal{D}}(\mathcal{M},-)$ commutes with direct limits.

PROPOSITION 3.1.6. *Assume that* $\mathcal{M}\in\text{Bimod}(\mathcal{C}-\mathcal{D})$ *and consider the following statements.*

1. $\mathcal{M}$ *is coherent.*
2. $-\otimes_{\mathcal{C}}\mathcal{M}$ *preserves finitely presented objects in* $\mathcal{C}$.

Then (1) implies (2). The converse holds if $\mathcal{C}$ *is generated by finitely presented objects.*

PROOF.

(1)⇒(2) Assume that T is a finitely presented object in $\mathcal{C}$. I.e. $\operatorname{Hom}_{\mathcal{C}}(T,-)$ commutes with direct limits. We have to show that $\operatorname{Hom}_{\mathcal{C}}(T\otimes_{\mathcal{C}}\mathcal{M},-)$ commutes with direct limits. This follows from the fact that

$$\operatorname{Hom}_{\mathcal{C}}(T\otimes_{\mathcal{C}}\mathcal{M},-)=\operatorname{Hom}_{\mathcal{C}}(T,\mathcal{H}om_{\mathcal{D}}(\mathcal{M},-))$$

(2)⇒(1) We need a natural isomorphism between $\operatorname{Hom}_{\mathcal{C}}(U,\varinjlim_i \operatorname{Hom}_{\mathcal{D}}(\mathcal{M},\mathcal{N}_i))$ and $\operatorname{Hom}_{\mathcal{C}}(U,\operatorname{Hom}_{\mathcal{D}}(\mathcal{M},\varinjlim \mathcal{N}_i))$ for an arbitrary inverse system $(\mathcal{N}_i)_i$ in $\mathcal{D}$ and $U\in\mathcal{C}$. Since $\mathcal{C}$ is generated by finitely presented objects, it suffices to construct this for U finitely presented. But this is clear by adjointness. □

PROPOSITION 3.1.7. *Assume $\mathcal{D}$ is a Grothendieck category. Suppose furthermore that $\mathcal{C}$ has exact direct limits. Let $\mathcal{M}\in$ BIMOD$(\mathcal{C}-\mathcal{D})$. Then the following are equivalent*

1. *$\mathcal{M}$ is coherent.*
2. *$\mathcal{H}om_{\mathcal{D}}(\mathcal{M},-)$ commutes with direct limits of injectives.*

PROOF. We only have to prove 2.⇒1. In a Grothendieck category embeddings into injectives can be constructed functorially [**17**, Thm 1.10.1]. Hence if $(T_i)_{i\in I}$ is an inverse system in $\mathcal{D}$ then there is a copresentation

$$0\to(T_i)_{i\in I}\to(E_i)_{i\in I}\to(F_i)_{i\in I}$$

with $(E_i)_i$, $(F_i)_i$ injective. The left exactness of $\mathcal{H}om_{\mathcal{D}}(\mathcal{M},-)$ together with the fact that direct limits are exact in $\mathcal{C}$ and $\mathcal{D}$ now shows what we want. □

COROLLARY 3.1.8. *Assume that $\mathcal{C}$ has exact direct limits and that $\mathcal{D}$ is locally noetherian. Then the category of coherent objects in* BIMOD$(\mathcal{C}-\mathcal{D})$ *is an abelian subcategory of* BIMOD$(\mathcal{C}-\mathcal{D})$*, closed under extensions.*

PROOF. According to Proposition 3.1.7 whether $\mathcal{M}\in$ BIMOD$(\mathcal{C}-\mathcal{D})$ is coherent can be tested on inverse systems of injectives $(E_i)_{i\in I}$. Since $\mathcal{D}$ is locally noetherian we also have that $F=\varinjlim_i E_i$ is injective. Hence by construction $\mathcal{H}om_{\mathcal{D}}(-,E_i)$ and $\mathcal{H}om_{\mathcal{D}}(-,F)$ are exact functors on BIMOD$(\mathcal{C}-\mathcal{D})$. The corollary is now a simple application of the five-lemma. □

Sometimes it is convenient to use "virtual" inverse limits of bimodules. These are defined below.

DEFINITION 3.1.9. Assume that $\mathcal{C}$ is cocomplete and that $(\mathcal{N}_n,\phi_{m,n})$ is an inverse system in BIMOD$(\mathcal{C}-\mathcal{D})$ indexed by $\mathbb{N}$. We define "$\varprojlim_n$"$\mathcal{N}_n$ by the rule

$$\mathcal{H}om_{\mathcal{D}}(\text{“}\varprojlim_n\text{”}\mathcal{N}_n,\mathcal{M})=\varinjlim_n \mathcal{H}om_{\mathcal{D}}(\mathcal{N}_n,\mathcal{M}) \tag{3.2}$$

for all $\mathcal{M}\in\mathcal{D}$. An inverse system $(\mathcal{N}_n,\phi_{m,n})$ such that "$\varprojlim_n$"$\mathcal{N}_n=0$ is called a *torsion inverse system*.

LEMMA 3.1.10. *Assume that $\mathcal{C}$ has exact direct limits and $\mathcal{D}$ has enough injectives. Then "$\varprojlim_n$" is exact. In particular the category of torsion inverse systems is closed under subobjects, quotients and extensions.*

PROOF. This is trivial. □

LEMMA 3.1.11. *Let* $(\mathcal{N}_n, \phi_{m,n})$ *be an inverse system in* $\mathcal{D}$ *(viewed as a subcategory of* $\mathrm{MOD}(\mathcal{D}) = \mathrm{BIMOD}(\mathbf{Ab} - \mathcal{D})$*). Then the following are equivalent*

1. $(\mathcal{N}_n, \phi_{m,n})$ *is torsion.*
2. *For every* m *there exists* $n \geq m$ *such that* $\phi_{m,n} : \mathcal{N}_n \to \mathcal{N}_m$ *is the zero morphism.*

PROOF. We prove (1) ⇒ (2) the other direction being obvious. We apply (3.2) with $\mathcal{M} = \mathcal{N}_m$. Then the identity morphism $\mathrm{id}_{\mathcal{N}_m}$ must become zero in some $\mathrm{Hom}_{\mathcal{D}}(\mathcal{N}_n, \mathcal{N}_m)$. This is exactly (2). □

REMARK 3.1.12. Note that in the previous lemma (1) ⇒ (2) holds more generally for inverse systems in $\mathrm{BIMOD}(\mathcal{C} - \mathcal{D})$.

If $C = (\mathcal{C}, \otimes, I)$ is a monoidal category [**24**] then an algebra object in $\mathcal{C}$ is a triple $(\mathcal{A}, \eta, \mu)$ where $\mathcal{A}$ is an object in $\mathcal{C}$ equipped with two morphisms $\eta : I \to \mathcal{A}$ (the unit) and $\mu : \mathcal{A} \otimes \mathcal{A} \to \mathcal{A}$ (the multiplication) satisfying the usual compatibilities.

It is clear that $(\mathrm{BIMOD}(\mathcal{D} - \mathcal{D}), \otimes_{\mathcal{D}}, \mathrm{id}_{\mathcal{D}})$ and $(\mathrm{Bimod}(\mathcal{D} - \mathcal{D}), \otimes_{\mathcal{D}}, \mathrm{id}_{\mathcal{D}})$ are monoidal categories so we denote the algebra objects in them respectively by $\mathrm{ALG}(\mathcal{D})$ and $\mathrm{Alg}(\mathcal{D})$. The objects in $\mathrm{ALG}(\mathcal{D})$ will be called *weak algebras* and those of $\mathrm{Alg}(\mathcal{D})$ will simply be called *algebras*. Furthermore we define $\mathrm{Mod}(\mathcal{A})$ as the category consisting of pairs $(\mathcal{M}, h)$ where $\mathcal{M} \in \mathcal{D}$ and h is a morphism $\mathcal{M} \otimes_{\mathcal{D}} \mathcal{A} \to \mathcal{M}$ in $\mathrm{MOD}(\mathcal{D})$ satisfying the usual compatibilities. Note that objects of $\mathrm{ALG}(\mathcal{D})$ are basically monads in the sense of [**24**] with some extra structure.

In order to interprete these definitions more concretely we remind the reader that $\mathrm{BIMOD}(\mathcal{D} - \mathcal{D}) = \mathcal{L}(\mathcal{D}, \mathcal{D})^{\mathrm{opp}}$. Thus $\mathrm{ALG}(\mathcal{D})$ is equivalent to the category of *coalgebra* objects in $\mathcal{L}(\mathcal{D}, \mathcal{D})$. Thus if $\mathcal{A} = (\mathcal{A}, \eta, \mu) \in \mathrm{ALG}(\mathcal{D})$ then the unit η is in fact a natural transformation

$$\eta : \mathcal{H}om_{\mathcal{D}}(\mathcal{A}, -) \to \mathrm{id}_{\mathcal{D}}$$

and the multiplication μ is a natural transformation

$$\mu : \mathcal{H}om_{\mathcal{D}}(\mathcal{A}, -) \to \mathcal{H}om_{\mathcal{D}}(\mathcal{A}, \mathcal{H}om_{\mathcal{D}}(\mathcal{A}, -)) \tag{3.3}$$

Likewise if $\mathcal{M} = (\mathcal{M}, h) \in \mathrm{Mod}(\mathcal{A})$ then h is a natural transformation

$$h : \mathrm{Hom}_{\mathcal{D}}(\mathcal{M}, -) \to \mathrm{Hom}_{\mathcal{D}}(\mathcal{M}, \mathcal{H}om_{\mathcal{D}}(\mathcal{A}, -)) \tag{3.4}$$

Then $\bar{h} = h_{\mathcal{M}}(\mathrm{id}_{\mathcal{M}})$ defines a morphism

$$\bar{h} : \mathcal{M} \to \mathcal{H}om_{\mathcal{D}}(\mathcal{A}, \mathcal{M}) \tag{3.5}$$

Conversely if one is given a morphism $\bar{h} : \mathcal{M} \to \mathcal{H}om_{\mathcal{D}}(\mathcal{A}, \mathcal{M})$ as in (3.5) then the composition

$$\mathrm{Hom}_{\mathcal{D}}(\mathcal{M}, -) \to \mathrm{Hom}_{\mathcal{D}}(\mathcal{H}om_{\mathcal{D}}(\mathcal{A}, \mathcal{M}), \mathcal{H}om_{\mathcal{D}}(\mathcal{A}, -))$$

with $\bar{h}^*$ is a natural transformation as in (3.4).

Elaborating on this one obtains the following results :

PROPOSITION 3.1.13. *Let* $\mathcal{A} \in \mathrm{ALG}(\mathcal{D})$. *Then* $(\mathcal{M}, h) \mapsto (\mathcal{M}, \bar{h})$ *defines an isomorphism between* $\mathrm{Mod}(\mathcal{A})$ *and the category of* $\mathcal{A}$*-comodules where we consider* $\mathcal{A}$ *as a coalgebra in* $\mathcal{L}(\mathcal{D}, \mathcal{D})$.

LEMMA 3.1.14. *Let $\mathcal{A} \in \mathrm{ALG}(\mathcal{D})$. The forgetful functor*

$$(-)_{\mathcal{D}} : \mathrm{Mod}(\mathcal{A}) \to \mathcal{D} : (\mathcal{M}, h) \mapsto \mathcal{M} \tag{3.6}$$

has a right adjoint given by $\mathcal{H}om_{\mathcal{D}}(\mathcal{A}, -)$ (with its canonical $\mathcal{A}$-structure given by (3.3)) and furthermore if $\mathcal{A} \in \mathrm{Alg}(\mathcal{D})$ then (3.6) also has a left adjoint given by $- \otimes_{\mathcal{D}} \mathcal{A}$.

$\mathrm{Mod}(\mathcal{A})$ inherits most of the good properties of $\mathcal{D}$, as is shown in the following proposition.

PROPOSITION 3.1.15. *Let $\mathcal{A} = (\mathcal{A}, \eta, \mu) \in \mathrm{ALG}(\mathcal{D})$.*

1. *$\mathrm{Mod}(\mathcal{A})$ is an abelian category.*
2. *$\mathrm{Mod}(\mathcal{A})$ possesses all colimits which exist in $\mathcal{D}$.*
3. *The forgetful functor $(-)_{\mathcal{D}}$ is exact, faithful and commutes with colimits.*
4. *$\mathrm{Mod}(\mathcal{A})$ is cogenerated by objects of the form $\mathcal{H}om_{\mathcal{D}}(\mathcal{A}, \mathcal{M})$, $\mathcal{M} \in \mathcal{D}$.*
5. *If $E \in \mathrm{Inj}(\mathcal{D})$ then $\mathcal{H}om_{\mathcal{D}}(\mathcal{A}, E) \in \mathrm{Inj}(\mathrm{Mod}(\mathcal{A}))$. In particular if $\mathcal{D}$ has enough injectives then so does $\mathrm{Mod}(\mathcal{A})$.*
6. *If $\mathcal{D}$ has exact direct limits then so does $\mathrm{Mod}(\mathcal{A})$.*

Assume now in addition that $\mathcal{A} \in \mathrm{Alg}(\mathcal{D})$

7. *$\mathrm{Mod}(\mathcal{A})$ possesses all limits that exist in $\mathcal{D}$ and $(-)_{\mathcal{D}}$ commutes with these limits.*
8. *$\mathrm{Mod}(\mathcal{A})$ is generated by objects of the form $\mathcal{M} \otimes_{\mathcal{D}} \mathcal{A}$, $\mathcal{M} \in \mathcal{D}$. Hence if $\mathcal{D}$ has a generator then so does $\mathrm{Mod}(\mathcal{A})$.*

In particular combining (6),(8) we find that if $\mathcal{A} \in \mathrm{Alg}(\mathcal{D})$ and $\mathcal{D}$ is a Grothendieck category then so is $\mathrm{Mod}(\mathcal{A})$.

PROOF. 2. Let I be a small category and $\mathcal{M} : I \to \mathrm{Mod}(\mathcal{A})$ a functor. We write $\mathcal{M}(i) = (\mathcal{M}_i, h_i)$. Then one easily verifies that $\varinjlim \mathcal{M}$ is given by the pair $(\varinjlim \mathcal{M}_i, h)$ where $\bar{h}$ is given by the composition

$$\varinjlim \mathcal{M}_i \xrightarrow{\varinjlim \bar{h}_i} \varinjlim \mathcal{H}om_{\mathcal{D}}(\mathcal{A}, \mathcal{M}_i) \xrightarrow{\text{can}} \mathcal{H}om_{\mathcal{D}}(\mathcal{A}, \varinjlim \mathcal{M}_i)$$

1. From (2) it follows that $\mathrm{Mod}(\mathcal{A})$ has cokernels. So we have to show that $\mathrm{Mod}(\mathcal{A})$ has kernels. Let $f : (\mathcal{M}, h) \to (\mathcal{N}, j)$ be a morphism in $\mathrm{Mod}(\mathcal{A})$. Then $\ker f$ is the pair $(\mathcal{K}, s)$ where $\mathcal{K} = \ker(\mathcal{M} \to \mathcal{N})$ and $\bar{s}$ is as in the following commutative diagram with exact rows.

$$\begin{array}{ccccccc}
0 \longrightarrow & \mathcal{K} & \longrightarrow & \mathcal{M} & \xrightarrow{f} & \mathcal{N} \\
& \bar{s}\downarrow & & \bar{h}\downarrow & & \bar{j}\downarrow \\
0 \longrightarrow & \mathcal{H}om_{\mathcal{D}}(\mathcal{A}, \mathcal{K}) & \longrightarrow & \mathcal{H}om_{\mathcal{D}}(\mathcal{A}, \mathcal{M}) & \xrightarrow{f} & \mathcal{H}om_{\mathcal{D}}(\mathcal{A}, \mathcal{N})
\end{array}$$

3. This follows from the explicit constructions of kernels and colimits in (1) and (2).
4. Assume that $(\mathcal{M}, h) \in \mathrm{Mod}(\mathcal{A})$. The composition

$$\mathcal{M} \xrightarrow{\bar{h}} \mathcal{H}om_{\mathcal{D}}(\mathcal{A}, \mathcal{M}) \xrightarrow{\eta} \mathcal{M}$$

is the identity so $\bar{h}$ is a monomorphism. We know already that $\mathcal{H}om_{\mathcal{D}}(\mathcal{A}, \mathcal{M})$ has a canonical structure as $\mathcal{A}$-module and it is easy to see that $\bar{h}$ is compatible with it.

5. This follows from the fact that $\mathcal{H}om_{\mathcal{D}}(\mathcal{A}, -)$ has a left adjoint which is exact by (3).
6. This follows from the explicit construction of kernels, cokernels and colimits in (1)(2).
7. Let $\mathcal{M} : I \to \mathrm{Mod}(\mathcal{A})$ be as in (2) Then $\varprojlim \mathcal{M}$ is the pair $(\varprojlim \mathcal{M}_i, h')$ where $\bar{h}'$ is the composition

$$\varprojlim \mathcal{M}_i \xrightarrow{\varprojlim \bar{h}_i} \varprojlim \mathcal{H}om_{\mathcal{D}}(\mathcal{A}, \mathcal{M}_i) \xrightarrow{\mathrm{can}^{-1}} \mathcal{H}om_{\mathcal{D}}(\mathcal{A}, \varprojlim \mathcal{M}_i)$$

where we have used the fact that $\mathcal{H}om_{\mathcal{D}}(\mathcal{A}, -)$ preserves products.
8. If $(\mathcal{M}, h) \in \mathrm{Mod}(\mathcal{A})$ then $\mathcal{M} \otimes_{\mathcal{D}} \mathcal{A} \in \mathrm{Mod}(\mathcal{A})$ by lemma 3.1.14. Furthermore the composition

$$\mathcal{M} \xrightarrow{\eta} \mathcal{M} \otimes_{\mathcal{D}} \mathcal{A} \xrightarrow{h} \mathcal{M}$$

is the identity so $\mathcal{M} \otimes_{\mathcal{D}} \mathcal{A} \to \mathcal{M}$ is an epimorphism. □

REMARK 3.1.16. We have no example where $\mathcal{D}$ is a Grothendieck category and $\mathrm{Mod}(\mathcal{A})$ is not a Grothendieck category (even if we only assume $\mathcal{A} \in \mathrm{ALG}(\mathcal{D})$).

Assume now that $f : \mathcal{A} \to \mathcal{B}$ is a morphism in $\mathrm{ALG}(\mathcal{D})$. Let $\mathcal{M} = (\mathcal{M}, h) \in \mathrm{Mod}(\mathcal{A})$, $\mathcal{N} = (\mathcal{N}, j) \in \mathrm{Mod}(\mathcal{B})$. Then we define $\mathcal{N}_{\mathcal{A}} \in \mathrm{Mod}(\mathcal{A})$ as the pair $(\mathcal{N}, j')$ were $\bar{j}'$ is the composition

$$\mathcal{N} \xrightarrow{\bar{j}} \mathcal{H}om_{\mathcal{D}}(\mathcal{B}, \mathcal{N}) \xrightarrow{f^*} \mathcal{H}om_{\mathcal{D}}(\mathcal{A}, \mathcal{N})$$

$\mathcal{H}om_{\mathcal{A}}(\mathcal{B}, \mathcal{M}) \in \mathrm{Mod}(\mathcal{B})$ is the pair $(\mathcal{U}, u)$ where $\mathcal{U}$ is obtained as the equalizer in $\mathrm{Mod}(X)$ of

$$\begin{array}{ccc} \mathcal{H}om_{\mathcal{D}}(\mathcal{B}, \mathcal{M}) & \xrightarrow{\mathrm{mult}^*} & \mathcal{H}om_{\mathcal{D}}(\mathcal{B} \otimes_{\mathcal{D}} \mathcal{B}, \mathcal{M}) \\ {\scriptstyle \bar{h}}\downarrow & & \downarrow{\scriptstyle (1\otimes f)^*} \\ \mathcal{H}om_{\mathcal{D}}(\mathcal{B}, \mathcal{H}om_{\mathcal{D}}(\mathcal{A}, \mathcal{M})) & = & \mathcal{H}om_{\mathcal{D}}(\mathcal{B} \otimes_{\mathcal{D}} \mathcal{A}, \mathcal{M}) \end{array} \tag{3.7}$$

and $\bar{u}$ is obtained from the fact that all objects in (3.7) carry a $\mathcal{B}$-structure, and the morphisms are compatible with it.

Finally if $\mathcal{A}, \mathcal{B} \in \mathrm{Alg}(\mathcal{D})$ then $\mathcal{M} \otimes_{\mathcal{A}} \mathcal{B}$ is defined as the coequalizer in $\mathrm{Mod}(X)$ of

$$\mathcal{M} \otimes_{\mathcal{D}} \mathcal{A} \otimes_{\mathcal{D}} \mathcal{B} \underset{(1\otimes \mathrm{mult})\circ(1\otimes f \otimes 1)}{\overset{\bar{h}\otimes 1}{\rightrightarrows}} \mathcal{M} \otimes_{\mathcal{D}} \mathcal{B}$$

So, summarizing, we have constructed the standard functors

$$(-)_{\mathcal{A}} : \mathrm{Mod}(\mathcal{B}) \to \mathrm{Mod}(\mathcal{A}) \tag{3.8}$$

$$\mathcal{H}om_{\mathcal{A}}(\mathcal{B}, -) : \mathrm{Mod}(\mathcal{A}) \to \mathrm{Mod}(\mathcal{B}) \tag{3.9}$$

and if $\mathcal{A}, \mathcal{B} \in \mathrm{Alg}(\mathcal{D})$

$$- \otimes_{\mathcal{A}} \mathcal{B} : \mathrm{Mod}(\mathcal{A}) \to \mathrm{Mod}(\mathcal{B}) \tag{3.10}$$

In general (3.9) is a right adjoint to (3.8) and if (3.10) is defined then it is a left adjoint to (3.8). From the constructions of limits, colimits, kernels and cokernels in (the proof of) Proposition 3.1.15 one verifies that (3.8) (3.9)(3.10) have the standard exactness properties and satisfy the usual compatibilities with (co)limits.

For $\mathcal{A},\mathcal{B}\in \mathrm{ALG}(\mathcal{D})$ we define

$$\mathrm{BIMOD}(\mathcal{A}-\mathcal{B}) = \mathrm{BIMOD}(\mathrm{Mod}(\mathcal{A})-\mathrm{Mod}(\mathcal{B}))$$
$$\mathrm{Bimod}(\mathcal{A}-\mathcal{B}) = \mathrm{Bimod}(\mathrm{Mod}(\mathcal{A})-\mathrm{Mod}(\mathcal{B}))$$

We denote $-\otimes_{\mathrm{Mod}(\mathcal{B})}-$ by $-\otimes_{\mathcal{B}}-$. In general we will replace in our notations $\mathrm{Mod}(\mathcal{X})$ by $\mathcal{X}$ when no confusion can arise.

Assume that $\mathcal{A}\to\mathcal{A}'$, $\mathcal{B}\to\mathcal{B}'$ are morphisms in $\mathrm{ALG}(\mathcal{D})$. We will now define functors

$$\mathcal{A}'\otimes_{\mathcal{A}}-\otimes_{\mathcal{B}}\mathcal{B}' : \mathrm{BIMOD}(\mathcal{A}-\mathcal{B})\to\mathrm{BIMOD}(\mathcal{A}'-\mathcal{B}') \tag{3.11}$$
$${}_{\mathcal{A}}(-)_{\mathcal{B}} : \mathrm{BIMOD}(\mathcal{A}'-\mathcal{B}')\to\mathrm{BIMOD}(\mathcal{A}-\mathcal{B}) \tag{3.12}$$

as we did for modules.

Let $\mathcal{M}\in\mathrm{BIMOD}(\mathcal{A}-\mathcal{B})$, $\mathcal{N}\in\mathrm{BIMOD}(\mathcal{A}'-\mathcal{B}')$. Then we define

$$\mathcal{H}om_{\mathcal{B}'}(\mathcal{A}'\otimes_{\mathcal{A}}\mathcal{M}\otimes_{\mathcal{B}}\mathcal{B}',-)\overset{\text{def}}{=}\mathcal{H}om_{\mathcal{A}}(\mathcal{A}',\mathcal{H}om_{\mathcal{B}}(\mathcal{M},(-)_{\mathcal{B}}))$$

$$\mathcal{H}om_{\mathcal{B}}({}_{\mathcal{A}}\mathcal{N}_{\mathcal{B}},-)\overset{\text{def}}{=}\mathcal{H}om_{\mathcal{B}'}(\mathcal{N},\mathcal{H}om_{\mathcal{B}}(\mathcal{B}',-))_{\mathcal{A}} \tag{3.13}$$

The functor $-\otimes_{\mathcal{A}'}\mathcal{A}'\otimes_{\mathcal{A}}\mathcal{M}\otimes_{\mathcal{B}}\mathcal{B}'$ should be given by

$$(-)_{\mathcal{A}}\otimes_{\mathcal{A}}\mathcal{M}\otimes_{\mathcal{B}}\mathcal{B}'$$

and the functor $-\otimes_{\mathcal{A}}{}_{\mathcal{A}}\mathcal{N}_{\mathcal{B}}$ should be given by

$$((-\otimes_{\mathcal{A}}\mathcal{A}')\otimes_{\mathcal{A}'}\mathcal{N})_{\mathcal{B}} \tag{3.14}$$

So we conclude that if $\mathcal{M}\in\mathrm{Bimod}(\mathcal{A}-\mathcal{B})$ and $\mathcal{N}\in\mathrm{Bimod}(\mathcal{A}'-\mathcal{B}')$ then if $\mathcal{B}'\in\mathrm{Alg}(\mathcal{D})$ then (3.11) respects "Bimod" and if $\mathcal{A}'\in\mathrm{Alg}(\mathcal{D})$ then (3.12) respects "Bimod".

Our definition of $\mathrm{BIMOD}(\mathcal{A}-\mathcal{B})$ has the advantage that we can directly apply Proposition 3.1.1 to obtain the properties of this category. However we would also like to have a definition which resembles more closely that of modules. Therefore we state the following proposition.

PROPOSITION 3.1.17. *The following categories are equivalent.*

1. $\mathrm{BIMOD}(\mathcal{A}-\mathcal{B})$
2. *The category of triples* $(\mathcal{M},h,h')$ *where* $\mathcal{M}\in\mathrm{BIMOD}(\mathcal{D}-\mathcal{D})$ *and* $h:\mathcal{A}\otimes_{\mathcal{D}}\mathcal{M}\to\mathcal{M}$, $h':\mathcal{M}\otimes_{\mathcal{D}}\mathcal{B}\to\mathcal{M}$ *are morphisms in* $\mathrm{BIMOD}(\mathcal{D}-\mathcal{D})$ *satisfying the usual compatibilities.*

If $\mathcal{A},\mathcal{B}\in\mathrm{Alg}(\mathcal{D})$ *then in the previous statement, "*BIMOD*" may be replaced by "*Bimod*".*

PROOF. This is a rather tedious verification which we leave to the reader. Let us simply state how one associates a left exact functor $\mathcal{H}om_{\mathcal{B}}(\mathcal{M},-):\mathrm{Mod}(\mathcal{B})\to\mathrm{Mod}(\mathcal{A})$ to a triple $(\mathcal{M},h,h')$.

Let $\mathcal{N}=(\mathcal{N},p)\in\mathrm{Mod}(\mathcal{B})$. Then $\mathcal{H}om_{\mathcal{B}}(\mathcal{M},\mathcal{N})$ will be the equalizer of

$$\begin{array}{ccc}\mathcal{H}om_{\mathcal{D}}(\mathcal{M},\mathcal{N}) & = & \mathcal{H}om_{\mathcal{D}}(\mathcal{M},\mathcal{N})\\ {\scriptstyle h'^*}\downarrow & & \downarrow{\scriptstyle \bar{p}}\\ \mathcal{H}om_{\mathcal{D}}(\mathcal{M}\otimes_{\mathcal{D}}\mathcal{B},\mathcal{N}) & = & \mathcal{H}om_{\mathcal{D}}(\mathcal{M},\mathcal{H}om_{\mathcal{D}}(\mathcal{B},\mathcal{N}))\end{array} \tag{3.15}$$

The $\mathcal{A}$-structure on $\mathcal{H}om_{\mathcal{B}}(\mathcal{M},\mathcal{N})$ is obtained from the fact that all objects in (3.15) carry a canonical $\mathcal{A}$-structure and the morphisms are compatible with it. □

Now let $\mathcal{A} \to \mathcal{B}$ be a morphism in ALG($\mathcal{D}$). From (3.11) we obtain a functor

$$- \otimes_{\mathcal{A}} \mathcal{B} : \mathrm{MOD}(\mathcal{A}) \to \mathrm{MOD}(\mathcal{B})$$

Assume that Mod($\mathcal{A}$) and Mod($\mathcal{B}$) have enough injectives. Then we define associated functors

$$\mathcal{T}or_i^{\mathcal{A}}(-, \mathcal{B}) : \mathrm{MOD}(\mathcal{A}) \to \mathrm{MOD}(\mathcal{B})$$

by

$$\mathrm{Hom}_{\mathcal{B}}(\mathcal{T}or_i^{\mathcal{A}}(-, \mathcal{B}), E) = \mathrm{Ext}^i_{\mathcal{A}}(-, E_{\mathcal{A}})$$

for injectives $E \in \mathrm{Mod}(\mathcal{B})$. One easily verifies that

$$\mathcal{T}or_i^{\mathcal{A}}(-, \mathcal{B})_{\mathcal{A}} = \mathcal{T}or_i^{\mathcal{A}}(-, \mathcal{B}_{\mathcal{A}})$$

LEMMA 3.1.18. *Assume that $\mathcal{D}$ has colimits and that* Mod($\mathcal{B}$) *and* Mod($\mathcal{A}$) *are locally noetherian. Then $\mathcal{T}or_i^{\mathcal{A}}(-, \mathcal{B})$ sends coherent objects in* MOD($\mathcal{A}$) *to coherent objects in* MOD($\mathcal{B}$).

PROOF. This is a formal verification. We will give the proof as an illustration of how the various hypotheses are used.

Assume that $\mathcal{M} \in \mathrm{MOD}(\mathcal{A})$ is coherent and let $(E_i)_i$ be a directed system of injectives in Mod($\mathcal{B}$). By Proposition 3.1.7 we have to show that

$$\mathrm{Hom}_{\mathcal{B}}(\mathcal{T}or_i^{\mathcal{A}}(\mathcal{M}, \mathcal{B}), \varinjlim_i E_i) = \varinjlim_i \mathrm{Hom}_{\mathcal{B}}(\mathcal{T}or_i^{\mathcal{A}}(\mathcal{M}, \mathcal{B}), E_i)$$

(taking into account that **Ab** has exact direct limits).

From the fact that Mod($\mathcal{B}$) is locally noetherian it follows that $\varinjlim_i E_i$ is injective. Hence we have to show that

$$\mathrm{Ext}^i_{\mathcal{A}}(\mathcal{M}, (\varinjlim_i E_i)_{\mathcal{A}}) = \varinjlim_i \mathrm{Ext}^i_{\mathcal{A}}(\mathcal{M}, E_{i\mathcal{A}})$$

The fact that Mod($\mathcal{A}$) is locally noetherian and that $\mathcal{M}$ is coherent implies that the righthand side of this equation is equal to $\mathrm{Ext}^i_{\mathcal{A}}(\mathcal{M}, \varinjlim_i E_{i\mathcal{A}})$. So it remains to show that $(\varinjlim_i E_i)_{\mathcal{A}} = \varinjlim_i E_{i\mathcal{A}}$. This follows easily from the explicit construction of $\varinjlim$ in the proof of Proposition 3.1.15.2 and the definition of $(-)_{\mathcal{A}}$. □

3.2. Graded modules, bimodules and algebras

In this section we have stated all our definitions in the ungraded case but we will mainly need them in the graded case. Luckily the generalization is trivial.

If $\mathcal{D}$ is an abelian category then we denote by $\tilde{\mathcal{D}}$ the category of $\mathbb{Z}$-graded objects over $\mathcal{D}$. Thus by definition an object in $\tilde{\mathcal{D}}$ is a sequence of objects $(\mathcal{M}_n)_n$ in $\mathcal{D}$ and $\mathrm{Hom}_{\tilde{\mathcal{D}}}((\mathcal{M}_n)_n, (\mathcal{N}_n)_n) = \prod_n \mathrm{Hom}_{\mathcal{D}}(\mathcal{M}_n, \mathcal{N}_n)$. We identify an object $\mathcal{M} \in \mathcal{D}$ with the object of $\tilde{\mathcal{D}}$ which is $\mathcal{M}$ in degree zero and zero elsewhere. As usual $\tilde{\mathcal{D}}$ is equipped with the shiftfunctor $\mathcal{M} \mapsto \mathcal{M}(1)$ where $\mathcal{M}(1)_n = \mathcal{M}_{n+1}$. Of course $\mathcal{M}(m)$, $m \in \mathbb{Z}$ is defined similarly. To simplify the notation we will often write $\oplus_n \mathcal{M}_n$ for $(\mathcal{M}_n)_n$.

By analogy with the ungraded case we define

$$\mathrm{BIGR}(\mathcal{C} - \mathcal{D}) = \{\text{left exact functors } \tilde{\mathcal{D}} \to \tilde{\mathcal{C}} \text{ commuting with shift}\}^{\mathrm{opp}}$$

$$\mathrm{Bigr}(\mathcal{C}-\mathcal{D}) = \{\mathcal{M} \in \mathrm{BIGR}(\mathcal{C}-\mathcal{D}) \mid \mathcal{M} \text{ has a left adjoint}\}$$

and

$$\mathrm{GR}(\mathcal{C}) = \mathrm{BIGR}(\mathbf{Ab}-\mathcal{C})$$

Objects in $\mathrm{BIGR}(\mathcal{C}-\mathcal{D})$ will be called *graded weak $\mathcal{C}$-$\mathcal{D}$ bimodules* and objects in $\mathrm{Bigr}(\mathcal{C}-\mathcal{D})$ will be called *graded bimodules.*

We will denote the left exact functor corresponding to $\mathcal{M} \in \mathrm{BIGR}(\mathcal{C}-\mathcal{D})$ with $\underline{\mathcal{H}om}_{\mathcal{D}}(\mathcal{M},-)$ in order to avoid confusion with the ungraded case. Similar conventions will apply to the use of $\underline{\mathcal{E}xt}$, $\underline{\mathcal{T}or}$.

Note that if $\mathcal{M} \in \mathrm{BIGR}(\mathcal{C}-\mathcal{D})$ (resp. $\mathrm{Bigr}(\mathcal{C}-\mathcal{D})$) then the composition (for $n \in \mathbb{Z}$)

$$\mathcal{D} \hookrightarrow \tilde{\mathcal{D}} \xrightarrow{\mathcal{M}} \tilde{\mathcal{C}} \xrightarrow{\text{degree } -n} \mathcal{C}$$

defines objects $\mathcal{M}_n$ in $\mathrm{BIMOD}(\mathcal{C}-\mathcal{D})$ (resp. $\mathrm{Bimod}(\mathcal{C}-\mathcal{D})$) which determine $\mathcal{M}$. We write $\mathcal{M} = \oplus_n \mathcal{M}_n$.

Obviously $\mathrm{BIGR}(\mathcal{D}-\mathcal{D})$, $\mathrm{Bigr}(\mathcal{D}-\mathcal{D})$ are monoidal categories and we will denote the algebra objects in them by $\mathrm{GRALG}(\mathcal{D})$ resp. $\mathrm{Gralg}(\mathcal{D})$. Consistent with our earlier coventions we speak respectively of *weak algebras* and *algebras.* It is easy to see that $\mathcal{A} \in \mathrm{GRALG}(\mathcal{D})$ is of the form $\mathcal{A} = \oplus_n \mathcal{A}_n$ with multiplication morphisms $\mathcal{A}_m \otimes_{\mathcal{D}} \mathcal{A}_n \to \mathcal{A}_{m+n}$ and a unit $\mathrm{id}_{\mathcal{D}} \to \mathcal{A}_0$ satisfying the usual compatibilities.

If $\mathcal{A} \in \mathrm{GRALG}(\mathcal{D})$ then $\mathrm{Gr}(\mathcal{A})$ is defined as $\mathrm{Mod}(\mathcal{A})$ but with $\mathcal{D}$ replaced by $\tilde{\mathcal{D}}$. Finally if $\mathcal{A}, \mathcal{B} \in \mathrm{GRALG}(\mathcal{D})$ then

$$\mathrm{BIGR}(\mathcal{A}-\mathcal{B}) = \mathrm{BIGR}(\mathrm{Gr}(\mathcal{A})-\mathrm{Gr}(\mathcal{B}))$$
$$\mathrm{Bigr}(\mathcal{A}-\mathcal{B}) = \mathrm{Bigr}(\mathrm{Gr}(\mathcal{A})-\mathrm{Gr}(\mathcal{B}))$$

In the sequel we will freely use graded versions of ungraded results stated in this section.

3.3. Quotients of the identity functor

This section is related to [**28**].

First assume that we are in the following situation: $i_* : \mathcal{E} \to \mathcal{D}$ is an exact faithful functor between abelian categories which has a right adjoint $i^!$. Assume furthermore that the unit $i_* i^! \to \mathrm{id}_{\mathcal{D}}$ is monic on objects. Let $\mathcal{E}'$ be the essential image of $\mathcal{E}$ in $\mathcal{D}$ under i_*. Then the following holds:

LEMMA 3.3.1. 1. $i^! i_* = \mathrm{id}_{\mathcal{E}}$.
2. *i_* is fully faithful.*
Let $\mathcal{E}'$ be the essential image of $\mathcal{E}$ under i_*.
3. *If $\mathcal{M} \in \mathcal{D}$ then the (injective) counit map $i_* i^! \mathcal{M} \to \mathcal{M}$ is the maximal subobject of $\mathcal{M}$ which lies in $\mathcal{E}'$.*
4. *$\mathcal{E}'$ is closed under subquotients.*

Assume now that i_* has a left adjoint i^*. Then

5. $i^* i_* = \mathrm{id}_{\mathcal{E}}$.
6. *If $\mathcal{M} \in \mathcal{D}$ then the counit map $\mathcal{M} \to i_* i^* \mathcal{M}$ is surjective and it is the maximal quotient of $\mathcal{M}$ which lies in $\mathcal{E}'$.*

PROOF. 1. Let $\mathcal{N} \in \mathcal{E}$. From the theory of adjoint functors we know that the composition of the canonical maps

$$i_*\mathcal{N} \to i_*i^!i_*\mathcal{N} \to i_*\mathcal{N}$$

is the identity. By hypotheses the last map is monic. Hence both maps must be isomorphisms. The hypotheses imply that i_* reflects isomorphism so $\mathcal{N} \to i^!i_*\mathcal{N}$ is an isomorphism as well.

2. This is an immediate consequence of (1). Indeed if $\mathcal{M}, \mathcal{N} \in \mathcal{E}$ then

$$\mathrm{Hom}_{\mathcal{D}}(i_*\mathcal{M}, i_*\mathcal{N}) = \mathrm{Hom}_{\mathcal{E}}(\mathcal{M}, i^!i_*\mathcal{N}) = \mathrm{Hom}_{\mathcal{E}}(\mathcal{M}, \mathcal{N})$$

3. Let $\mathcal{M} \in \mathcal{D}$, $\mathcal{N}' \in \mathcal{E}'$. Then there is an object $\mathcal{N}$ in $\mathcal{E}$ such that $i_*\mathcal{N}$ is isomorphic to $\mathcal{N}'$. In particular there is a map $\phi : i_*\mathcal{N} \to \mathcal{M}$ whose image is $\mathcal{N}'$.

   Since

$$\mathrm{Hom}_{\mathcal{D}}(i_*\mathcal{N}, i_*i^!\mathcal{M}) = \mathrm{Hom}_{\mathcal{E}}(\mathcal{N}, i^!\mathcal{M}) = \mathrm{Hom}_{\mathcal{D}}(i_*\mathcal{N}, \mathcal{M})$$

   we find that ϕ factors through $i_*i^!\mathcal{M}$. Thus $\mathcal{N}' \subset i_*i^!\mathcal{M}$.

4. Since i_* is left exact $\mathcal{E}'$ is closed under kernels. So it suffices to show that $\mathcal{E}'$ is closed under quotients. Let $\mathcal{N} \in \mathcal{E}'$ and let $\mathcal{N}'$ be a quotient of $\mathcal{N}$. Then we have a commutative diagram

$$\begin{array}{ccc} i_*i^!\mathcal{N} & \xrightarrow{\alpha} & \mathcal{N} \\ \downarrow & & \downarrow \beta \\ i_*i^!\mathcal{N}' & \xrightarrow[\gamma]{} & \mathcal{N}' \end{array}$$

   where α is an isomorphism, β is surjective and γ is injective. This implies that γ is an isomorphism.

5. This is proved in a similar way as (1).
6. This is proved in a similar way as (3). □

Let us call a full subcategory of an abelian category *weakly closed* if it is closed under subquotients and if the inclusion functor has a right adjoint. In most cases this is equivalent to the definition of a closed subcategory in [**16**, IV.4]. We will call a weakly closed subcategory *closed* if the inclusion functor also has a left adjoint.

Lemma 3.3.1 tells us that $\mathcal{E}'$ is a weakly closed subcategory of $\mathcal{D}$ and if i^* exists then $\mathcal{E}'$ is even closed. Conversely if $\mathcal{E}$ is a (weakly) closed subcategory of $\mathcal{D}$ and i_* is the inclusion functor then i_* satisfies the hypotheses of Lemma 3.3.1.

Assume now that $\mathcal{D}$, $\mathcal{E}$ have enough injectives so that the weak bimodule categories are abelian (cf Prop. 3.1.1). Let $\mathcal{B}$ be a quotient of $\mathrm{id}_{\mathcal{D}}$ in $\mathrm{ALG}(\mathcal{D})$ (by this we mean that the underlying $\mathcal{D}$-bimodule map is an epimorphism).

We will denote the functor $(-)_{\mathcal{D}}$ by i_* and its right adjoint $\mathcal{H}om_{\mathcal{D}}(\mathcal{B}, -)$ by $i^!$. If $\mathcal{B} \in \mathrm{Alg}(\mathcal{D})$ then we denote $- \otimes_{\mathcal{D}} \mathcal{B}$ by i^*. This is the left adjoint to i_*.

LEMMA 3.3.2. *The functors i_*, $i^!$ and i^* (if $\mathcal{B} \in \mathrm{Alg}(\mathcal{D})$) satisfy the hypotheses of lemma 3.3.1.*

PROOF. That i_* is exact and faithful is Proposition 3.1.15.3. To check the other hypothesis we note the easily verified fact that the counit map for $(i_*, i^!)$ is the composition

$$i_*i^!\mathcal{M} = \mathcal{H}om_{\mathcal{D}}(\mathcal{B}, \mathcal{M}) \hookrightarrow \mathcal{H}om_{\mathcal{D}}(\mathrm{id}_{\mathcal{D}}, \mathcal{M}) = \mathcal{M}$$

Hence it is monic by Proposition 3.1.2.1.

□

Using a slight abuse of notation we will write $i_* \operatorname{Mod}(\mathcal{B})$ for the essential image of $\operatorname{Mod}(\mathcal{B})$ in $\mathcal{D}$. Lemma 3.3.1 basically tells us that the functor $\mathcal{B} \to i_* \operatorname{Mod}(\mathcal{B})$ associates with every quotient of $\mathrm{id}_{\mathcal{D}}$ a weakly closed subcategory of $\mathcal{D}$ and if the quotient is in $\operatorname{Alg}(\mathcal{D})$ then the resulting category is closed. We will now show that the converse to this essentially holds.

Assume that $\mathcal{E} \subset \mathcal{D}$ is a weakly closed subcategory. Let i_* be the inclusion functor and $i^!$ its right adjoint. Let $\mathcal{B}$ be the left exact functor $i_* i^!$. Then the counit $i_* i^! \to \mathrm{id}_{\mathcal{D}}$ and the comultiplication $i_* i^! = i_* \mathrm{id}_{\mathcal{E}} i^! \to i_* i^! i_* i^!$ make $\mathcal{B}$ into a weak algebra (remember that a "weak algebra" is actually a coalgebra object in the the monoidal category of left exact functors, see the discussion after Prop. 3.1.5). Furthermore the counit $i_* i^! \to \mathrm{id}_{\mathcal{D}}$ makes $\mathcal{B}$ into a quotient of $\mathrm{id}_{\mathcal{D}}$ (again remember that arrows are reversed if we pass from functors to bimodules).

Now by a routine verification (or using an appropriate version of Beck's theorem [**24**]) one shows that $\operatorname{Mod}(\mathcal{B}) \cong \mathcal{E}$. Elaborating on this one may prove the following result.

PROPOSITION 3.3.3. *With notations as above. The functors $\mathcal{B} \mapsto i_* \operatorname{Mod}(\mathcal{B})$ and $\mathcal{E} \mapsto i_* i^!$ induce inverse bijections between the quotients of $\mathrm{id}_{\mathcal{D}}$ in $\operatorname{ALG}(\mathcal{D})$ (resp. $\operatorname{Alg}(\mathcal{D})$) and the weakly closed (resp. closed) subcategories in $\mathcal{D}$.*

Below we state a few elementary results concerning weakly closed and closed categories which will be useful in the sequel. Recall that a category is said to be well-powered if the set of subobjects of an arbitrary object is small. In an abelian category this holds if there is a generator or a cogenerator. Hence in particular a Grothendieck category is well-powered.

PROPOSITION 3.3.4. *Let $\mathcal{E}$ be a full subcategory of the abelian category $\mathcal{D}$.*

1. *If $\mathcal{E}$ is weakly closed in $\mathcal{D}$ and if $\mathcal{D}$ is cocomplete, then so is $\mathcal{E}$.*
2. *If $\mathcal{D}$ is cocomplete and well-powered and if $\mathcal{E}$ is closed under subquotients and direct sums (in $\mathcal{D}$) then $\mathcal{E}$ is weakly closed.*
3. *If $\mathcal{E}$ is closed and if $\mathcal{D}$ is complete then so is $\mathcal{E}$.*
4. *If $\mathcal{D}$ is complete and well-powered and if $\mathcal{E}$ is weakly closed and is closed under products (in $\mathcal{D}$) then $\mathcal{E}$ is closed.*

PROOF. Let $i_* : \mathcal{E} \to \mathcal{D}$ be the inclusion functor and let $i^!$, i^* be the right and left adjoint to i_* if they exist.

1. It is sufficient to show that $\mathcal{E}$ is closed under direct sums. Let $\oplus_{j \in J} M_j$ be such a direct sum. By construction $i^!(\oplus_{j \in J} M_j)$ is the maximal subobject of $\oplus_{j \in J} M_j$ contained in $\mathcal{E}$. Hence for all j, $M_j \subset i^!(\oplus_{j \in J} M_j)$. But this implies $i^!(\oplus_{j \in J} M_j) = \oplus_{j \in J} M_j$ and thus $\oplus_{j \in J} M_j \in \mathcal{E}$.
2. Since the set of subobjects of an object M is small and since $\mathcal{E}$ is closed under (small) direct unions, M has a largest subobject N which lies in $\mathcal{E}$. The assignment $M \mapsto N$ is the right adjoint to the inclusion functor.
3. This is proved by an argument dual to (1).
4. This is proved by an argument dual to (2). □

PROPOSITION 3.3.5. *If $\mathcal{D}$ is a weakly closed subcategory of $\mathcal{C}$ and $\mathcal{E}$ is a weakly closed subcategory of $\mathcal{D}$ then $\mathcal{E}$ is weakly closed in $\mathcal{C}$. A similar statement holds if we replace "weakly closed" by "closed".*

PROOF. This follows from the fact that adjoint functors are compatible with composition. □

Let us recall the definition of the *Gabriel product.* If $\mathcal{E}_1, \mathcal{E}_2$ are full subcategories of an abelian category $\mathcal{D}$ then $\mathcal{E}_1 \cdot \mathcal{E}_2$ is the full subcategory of $\mathcal{D}$ whose objects are given by middle terms of exact sequences

$$0 \to M_2 \to M \to M_1 \to 0$$

with $M_1 \in \mathcal{E}_1$, $M_2 \in \mathcal{E}_2$. It is easy to see that if $\mathcal{E}_1$, $\mathcal{E}_2$ are closed under subquotients then so is $\mathcal{E}_1 \cdot \mathcal{E}_2$.

PROPOSITION 3.3.6. **[28]** *If $\mathcal{E}_1, \mathcal{E}_2 \subset \mathcal{D}$ are (weakly) closed then so is $\mathcal{E}_1 \cdot \mathcal{E}_2$.*

PROOF. We give the proof for weak closedness. Closedness is similar. Let i_{1*}, i_{2*}, i_{12*} be the embeddings $\mathcal{E}_1 \subset \mathcal{D}$, $\mathcal{E}_2 \subset \mathcal{D}$, $\mathcal{E}_1 \cdot \mathcal{E}_2 \subset \mathcal{D}$. One has to construct the right adjoint to i_{12*}. Let $q : \mathcal{M} \to \mathcal{M}/i_2^!\mathcal{M}$ be the quotient map. Then one verifies that

$$i_{12}^!\mathcal{M} \stackrel{\text{def}}{=} q^{-1}(i_1^!(\mathcal{M}/i_2^!\mathcal{M}))$$

has the required properties. □

3.4. Ideals in the identity functor

In this section $\mathcal{C}$, $\mathcal{D}$, $\mathcal{E}$ will be abelian categories having enough injectives.

DEFINITION 3.4.1. If $\mathcal{A} \in \mathrm{ALG}(\mathcal{D})$ (resp. in $\mathrm{Alg}(\mathcal{D})$) then a weak ideal (resp. an ideal) in $\mathcal{A}$ is a subobject of ${}_{\mathcal{A}}\mathcal{A}_{\mathcal{A}}$ in $\mathrm{BIMOD}(\mathcal{A} - \mathcal{A})$ (resp. in $\mathrm{Bimod}(\mathcal{A} - \mathcal{A})$).

If I, J are weak ideals in $\mathcal{A} \in \mathrm{ALG}(\mathcal{D})$ then we define the weak ideal IJ as the image of the composition

$$I \otimes_{\mathcal{A}} J \to \mathcal{A} \otimes_{\mathcal{A}} \mathcal{A} = \mathcal{A}$$

If $f : \mathcal{A} \to \mathcal{B}$ is a morphism in $\mathrm{ALG}(\mathcal{D})$ then by $\ker f$ we denote $\ker({}_{\mathcal{D}}\mathcal{A}_{\mathcal{D}} \to {}_{\mathcal{D}}\mathcal{B}_{\mathcal{D}})$. Since $\ker f$ is canonically an object in $\mathrm{BIMOD}(\mathcal{A} - \mathcal{A})$ we find that $\ker f$ is a weak ideal in $\mathcal{A}$.

We will say that f is epic if the underlying bimodule map ${}_{\mathcal{D}}\mathcal{A}_{\mathcal{D}} \to {}_{\mathcal{D}}\mathcal{B}_{\mathcal{D}}$ is epic. If this is the case then we call the pair $(\mathcal{B}, f)$ a quotient object of $\mathcal{A}$ in $\mathrm{ALG}(\mathcal{D})$.

If I is a weak ideal in $\mathcal{A}$ then there is a unique weak algebra structure on $\mathcal{A}/I$ which makes the quotient map $q : \mathcal{A} \to \mathcal{A}/I$ into a morphism in $\mathrm{ALG}(\mathcal{D})$.

LEMMA 3.4.2. *$I \mapsto (\mathcal{A}/I, q)$ and $(\mathcal{B}, f) \mapsto \ker f$ define inverse bijections between*

1. *Subobject of ${}_{\mathcal{A}}\mathcal{A}_{\mathcal{A}}$ in* $\mathrm{BIMOD}(\mathcal{A} - \mathcal{A})$
2. *Isomorphism classes of quotient object of $\mathcal{A}$.*

PROOF. We leave this verification to the reader. The lemma can of course be proved more generally in the setting of monoidal categories $(\mathcal{C}, \otimes, I)$ where $\mathcal{C}$ is abelian and $\otimes$ is right exact. □

Recall that $\mathrm{BIMOD}(\mathcal{A} - \mathcal{A})$ was defined as $\mathrm{BIMOD}(\mathrm{MOD}(\mathcal{A}) - \mathrm{MOD}(\mathcal{A}))$ and similarly for "Bimod". Hence there is a 1-1 correspondence between (weak) ideals in $\mathcal{A}$ and (weak) ideals in $\mathrm{id}_{\mathrm{Mod}(\mathcal{A})}$ and furthermore this correspondence is compatible with products. Hence to simplify the notation below we will replace $\mathrm{Mod}(\mathcal{A})$ by $\mathcal{D}$ and $\mathcal{A}$ by $\mathrm{id}_{\mathcal{D}}$.

If $M \in \text{Mod}(\mathcal{D})$ and I is a weak ideal in $\text{id}_{\mathcal{D}}$ then we define M_I as the quotient object of M given by the image of $M \to \mathcal{H}om_{\mathcal{D}}(I, M)$. One quickly verifies :

LEMMA 3.4.3. 1. *$M \in \text{Mod}(\text{id}_{\mathcal{D}}/I)$ iff $M_I = 0$.*
2. *$(M_J)_I = M_{IJ}$.*

PROPOSITION 3.4.4. *Let I, J be weak ideals in $\text{id}_{\mathcal{D}}$. Then*

$$\text{Mod}(\text{id}_{\mathcal{D}}/IJ) = \text{Mod}(\text{id}_{\mathcal{D}}/I) \cdot \text{Mod}(\text{id}_{\mathcal{D}}/J)$$

PROOF. "$\subset$" Let $M \in \text{Mod}(\text{id}_{\mathcal{D}}/IJ)$. Then $M_{IJ} = 0$ and hence $(M_J)_I = 0$. Thus $M_J \in \text{Mod}(\text{id}_{\mathcal{D}}/I)$. Since by construction $\ker(M \to M_J) \in \text{Mod}(\text{id}_{\mathcal{D}}/J)$ we are done.

"$\supset$" Let $M \in \text{Mod}(\text{id}_{\mathcal{D}}/I) \cdot \text{Mod}(\text{id}_{\mathcal{D}}/J)$. Then there is an exact sequence

$$0 \to M_2 \to M \to M_1 \to 0$$

with $M_2 \in \text{Mod}(\text{id}_{\mathcal{D}}/J)$, $M_1 \in \text{Mod}(\text{id}_{\mathcal{D}}/I)$. In the commutative diagram

$$\begin{array}{ccc} M_2 & \longrightarrow & M \\ \downarrow & & \downarrow \\ (M_2)_J & \longrightarrow & M_J \end{array}$$

we have $(M_2)_J = 0$ and thus the composition $M_2 \to M \to M_J$ is zero. In other words, $M \to M_J$ factors through M_1. Thus M_J is a quotient of M_1 and hence $M_{IJ} = (M_J)_I = 0$. So $M \in \text{Mod}(\text{id}_{\mathcal{D}}/IJ)$. □

Our next aim is to characterize the ideals in $\text{id}_{\mathcal{D}}$ in terms of certain subcategories of $\mathcal{D}$. The following is a trivial consequence of Prop. 3.1.1(6) and Prop. 3.3.3.

PROPOSITION 3.4.5. *Assume that $\mathcal{D}$ is complete and has exact direct products and an injective cogenerator. Then a weak ideal in $\text{id}_{\mathcal{D}}$ is an ideal if and only if $\text{Mod}(\text{id}_{\mathcal{D}}/I)$ is closed.*

Hence our problems lie in the cases where products are not exact.

For simplicity we assume below that $\mathcal{D}$ is complete and has an injective cogenerator. Let us first recall the definition of the derived functors of the product functor. Assume that J is some index set and let $\prod_{j \in J} \mathcal{D}$ be the category whose objects consist of all families $(M_j)_{j \in J}$ with $M_j \in \mathcal{D}$. Then we denote by $\prod$ the functor

$$\prod : \prod_{j \in J} \mathcal{D} \to \mathcal{D} : (M_j)_{j \in J} \mapsto \prod_{j \in J} M_j$$

which is obviously left exact. We denote the derived functors of $\prod$ by $(R^i \prod)_i$. These functors are computed in the standard way. Let $(M_j)_{j \in J}$ be in $\prod_{j \in J} \mathcal{D}$. One takes injective resolutions $M_j \to E_j^{\cdot}$ and then one has

$$R^i \prod (M_j)_{j \in J} = H^i(\prod_{j \in J} E_j^{\cdot})$$

DEFINITION 3.4.6. A closed subcategory $\mathcal{E} \subset \mathcal{D}$ is called *well-closed* if for all families $(E_j)_{j \in J}$ whose members are injective in $\mathcal{E}$ (but not necessarily in $\mathcal{D}$) one has $R^1 \prod (E_j)_j = 0$.

PROPOSITION 3.4.7. *Let $\mathcal{D}$ be as above. That is, $\mathcal{D}$ is complete and has an injective cogenerator. Let $I \subset \mathrm{id}_{\mathcal{D}}$ be a weak ideal. Then I is an ideal if and only if* $\mathrm{Mod}(\mathrm{id}_{\mathcal{D}}/I)$ *is well-closed.*

PROOF. "$\Rightarrow$" Since I is an ideal, $\mathrm{id}_{\mathcal{D}}/I \in \mathrm{Alg}(\mathcal{D})$. Hence $\mathrm{Mod}(\mathrm{id}_{\mathcal{D}}/I)$ is certainly closed. Now let $(E_j)_{j\in J}$ be a family on injectives in $\mathcal{E}$ and let $(F_j)_{j\in J}$ be the corresponding injective hulls in $\mathcal{D}$, Since F_j is an essential extension of E_j and E_j is injective we have

$$\mathcal{H}om_{\mathcal{D}}(\mathrm{id}_{\mathcal{D}}/I, F_j) = E_j$$

Applying $\mathcal{H}om_{\mathcal{D}}(-, \prod_j F_j)$ to the exact sequence in $\mathrm{Bimod}(\mathcal{D} - \mathcal{D})$

$$0 \to I \to \mathrm{id}_{\mathcal{D}} \to \mathrm{id}_{\mathcal{D}}/I \to 0 \tag{3.16}$$

yields an exact sequence

$$0 \to \prod_j E_j \to \prod_j F_j \to \prod_j \mathcal{H}om_{\mathcal{D}}(I, F_j) \to 0 \tag{3.17}$$

which is obviously the direct product of the short exact sequence

$$0 \to E_j \to F_j \to \mathcal{H}om_{\mathcal{D}}(I, F_j) \to 0 \tag{3.18}$$

obtained from applying $\mathcal{H}om(-, F_j)$ to (3.16).

Now applying the long exact sequence for $(R^i \prod)_i$ to (3.18) yields that $R^1 \prod(E_j)_j = 0$.

"$\Leftarrow$" According to Prop. 3.1.1(5) it suffices to show that $\mathcal{H}om_{\mathcal{D}}(I, -)$ commutes with products, when evaluated on injectives. Let $(F_j)_j$ be a family of injectives in $\mathcal{D}$ and put $E_j = \mathcal{H}om_{\mathcal{D}}(\mathrm{id}_{\mathcal{D}}/I, F_j)$. We have again the short exact sequences given by (3.18) and using the fact that $R^1 \prod(E_j)_j = 0$ we obtain the short exact sequence (3.17) from the long exact sequence for $(R^i \prod)_i$.

On the other hand if we apply $\mathcal{H}om_{\mathcal{D}}(-, \prod(F_j)_j)$ to the exact sequence (3.16) we obtain

$$0 \to \prod_j E_j \to \prod_j F_j \to \mathcal{H}om_{\mathcal{D}}(I, \prod F_j) \to 0 \tag{3.19}$$

Comparing (3.17) and (3.19) yields what we want. □

Now we prove an analog for Proposition 3.3.5 for well-closed subcategories.

PROPOSITION 3.4.8. *Assume that $\mathcal{C}$ is complete and has an injective cogenerator. Assume that $\mathcal{D}$ is well-closed in $\mathcal{C}$ and that $\mathcal{E}$ is well-closed in $\mathcal{D}$. Then $\mathcal{E}$ is well-closed in $\mathcal{C}$.*

PROOF. Note that by Prop. 3.3.4 $\mathcal{D}$ is complete and it is clear that $\mathcal{D}$ has an injective cogenerator. So it makes sense to speak of a well-closed subcategory of $\mathcal{D}$.

We already know that $\mathcal{E}$ is closed in $\mathcal{C}$, so we only have to show that the derived product $R^1 \prod(E_\alpha)_\alpha$ is zero for a family of injectives in $\mathcal{E}$. Let

$$0 \to E_\alpha \to F_{\alpha 0} \to F_{\alpha 1} \to F_{\alpha 2}$$

be (truncated) injective resolutions of E_α in $\mathcal{D}$. Furthermore let

$$0 \to F_{\alpha i} \to G_{\alpha i 0} \to G_{\alpha i 1} \to G_{\alpha i 2}$$

be (truncated) Cartan Eilenberg resolutions for the complexes $F_\alpha^{\cdot}$.

Taking products this yields a diagram of complexes

$$\begin{array}{ccccc} & & & & 0 \\ & & & & \downarrow \\ 0 & \longrightarrow & \prod_\alpha E_\alpha & \longrightarrow & \prod_\alpha F^\cdot_\alpha \\ & & & & \downarrow \\ & & & & \prod_\alpha G^\cdot_\alpha \end{array}$$

By hypotheses the columns of this diagram are exact and so is the first row. Then it easily follows that

$$0 \to \prod_\alpha E_\alpha \to \prod_\alpha G_{\alpha 00} \to \prod_\alpha G_{\alpha 10} \oplus \prod_\alpha G_{\alpha 01} \to \prod_\alpha G_{\alpha 20} \oplus \prod_\alpha G_{\alpha 11} \oplus \prod_\alpha G_{\alpha 02}$$

is exact. Since this is a product of truncated injective resolutions of E_α in $\mathcal{C}$ we are done. □

We don't know if well-closedness is compatible with the Gabriel product and hence we don't know if the product of ideals is an ideal. In order to deal with this problem in the sequel we introduce one more technical notion.

DEFINITION 3.4.9. Assume that $\mathcal{D}$ is complete and has an injective cogenerator. A closed subcategory $\mathcal{E} \subset \mathcal{D}$ is *very well-closed* if for all families $(M_j)_{j\in J}$ of objects in $\mathcal{E}$ one has $R^1 \prod (M_j)_j = 0$.

PROPOSITION 3.4.10. *Assume that $\mathcal{C}$ is complete and has an injective cogenerator. Assume that $\mathcal{D}$ is well-closed in $\mathcal{C}$ and $\mathcal{E}$ is very well-closed in $\mathcal{D}$. Then $\mathcal{E}$ is very well-closed in $\mathcal{C}$.*

PROOF. This is proved similarly as Prop. 3.4.8, □

COROLLARY 3.4.11. *Assume that $\mathcal{D}$ is complete and has an injective cogenerator. Then a closed subcategory $\mathcal{E} \subset \mathcal{D}$ is very well-closed if and only if*

1. *$\mathcal{E}$ is well-closed.*
2. *$\mathcal{E}$ has exact direct products.*

PROOF. One direction is clear. To prove the other direction we use Proposition 3.4.10 with $\mathcal{E} = \mathcal{D}$. □

PROPOSITION 3.4.12. *Assume that $\mathcal{D}$ is complete and has an injective cogenerator. Let $\mathcal{E}_1, \mathcal{E}_2 \subset \mathcal{D}$ be very well-closed subcategories. Then $\mathcal{E}_1 \cdot \mathcal{E}_2$ is very well-closed.*

PROOF. Let $(M_j)_j$ be a family of objects in $\mathcal{E}_1 \cdot \mathcal{E}_2$. We have exact sequences

$$0 \to M_{j2} \to M_j \to M_{j1} \to 0$$

with $M_{ji} \in \mathcal{E}_i$. From the long exact sequence for $(R^i \prod)_i$ and the hypotheses we deduce that $R^1 \prod (M_j)_j = 0$. □

If $\mathcal{L} \in \text{BIMOD}(\mathcal{D} - \mathcal{D})$ is such that $\mathcal{H}om_{\mathcal{D}}(\mathcal{L}, -)$ is an equivalence of categories then we call $\mathcal{L}$ an *invertible* bimodule. Obviously in that case $\mathcal{L} \in \text{Bimod}(\mathcal{D} - \mathcal{D})$, and there exist $\mathcal{L}^{-1} \in \text{Bimod}(\mathcal{D} - \mathcal{D})$ such that $\mathcal{L} \otimes_{\mathcal{D}} \mathcal{L}^{-1} \cong \mathcal{L}^{-1} \otimes_{\mathcal{D}} \mathcal{L} \cong \text{id}_{\mathcal{D}}$.

Assume that $\mathcal{L}$ is invertible. We will call a weak ideal in $\mathcal{L}$ a subobject I of $\mathcal{L}$ in $\text{BIMOD}(\mathcal{D} - \mathcal{D})$. If I actually lies in $\text{Bimod}(\mathcal{D} - \mathcal{D})$ then we call I an ideal in $\mathcal{L}$.

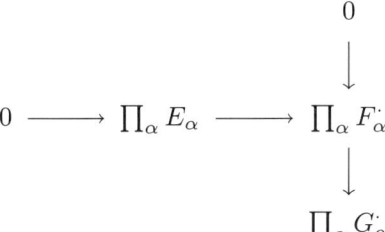

Clearly $I \mapsto I \otimes_{\mathcal{D}} \mathcal{L}^{-1}$ and $I \mapsto \mathcal{L}^{-1} \otimes_{\mathcal{D}} I$ induce bijections between (weak) ideals in $\mathcal{L}$ and (weak) ideals in $\mathrm{id}_{\mathcal{D}}$.

If $\mathcal{L}, \mathcal{M}$ are invertible $\mathcal{D} - \mathcal{D}$-bimodules and $I \subset \mathcal{L}$, $J \subset \mathcal{M}$ are weak ideals then we define IJ as the image of $I \otimes_{\mathcal{D}} J$ in $\mathcal{L} \otimes_{\mathcal{D}} \mathcal{M}$.

DEFINITION 3.4.13. Let $I \subset \mathcal{L}$ be a weak ideal in an invertible bimodule $\mathcal{L}$. Then the *Rees algebra* $\mathcal{D}(I)$ is the graded weak algebra given by

$$\mathrm{id}_{\mathcal{D}} \oplus I \oplus I^2 \oplus \cdots$$

(with obvious multiplication).

Clearly $\mathcal{D}(I) \in \mathrm{ALG}(\mathcal{D})$. However if $\mathrm{Mod}(\mathrm{id}_{\mathcal{D}}/I)$ is very well-closed then by Prop. 3.4.4 and 3.4.12 $\mathcal{D}(I)$ lies in $\mathrm{Alg}(\mathcal{D})$. It would be useful if we could find weaker conditions under which the Rees algebra of an ideal lies in $\mathrm{Alg}(\mathcal{D})$.

We close this section by introducing some terminology which we will use later.

If $I \subset \mathrm{id}_{\mathcal{D}}$ is a weak ideal defining a (weakly closed) subcategory $\mathcal{E}$ of $\mathcal{D}$ then $I/I^2 \in \mathrm{BIMOD}(\mathcal{E} - \mathcal{E})$. Following [**28**] we call I/I^2 the *conormal bundle* of $\mathcal{E}$ in $\mathcal{D}$.

3.5. Quasi-schemes

For us a quasi-scheme X is a Grothendieck category which we denote by $\mathrm{Mod}(X)$. A morphism $\alpha : Y \to X$ of quasi-schemes will be an additive functor $\alpha^* : \mathrm{Mod}(X) \to \mathrm{Mod}(Y)$ commuting with colimits. We will say that X is isomorphic to Y if $\mathrm{Mod}(X)$ is equivalent to $\mathrm{Mod}(Y)$. The quasi-schemes form a category which we denote by Qsch. Actually we will consider Qsch as a 2-category whose 2-cells correspond to natural isomorphisms (see Appendix A).

Commutative diagrams in a 2-category are usually only assumed to be commutative up to an explicit natural isomorphism (see Appendix A). Such diagrams are sometimes called pseudo-commutative diagrams, but we will call them simply "commutative diagrams". Likewise we will speak of "functors" when we actually mean pseudo-functors (again see Appendix A). A typical pseudo-functor is "Spec" which is defined below.

If $\alpha : Y \to X$ is a morphism of quasi-schemes then it follows from Theorem 2.1 that the adjoint to α^* exists. This adjoint is unique up to unique isomorphism and we will denote it by α_*. The assignment $\alpha \mapsto \alpha^*$ is obviously functorial since formally $\alpha = \alpha^*$! In addition, as is explained in Appendix A, the assignment $\alpha \mapsto \alpha_*$ is a pseudo-functor from the 2-category quasi-schemes to the 2-category of the category of Grotendieck categories.

Denote by "Sch" the category of quasi-compact, quasi-separated schemes. If $X \in \mathrm{Sch}$ then the category of quasi-coherent sheaves on X is a Grothendieck category [**32**]. In that case we define $\mathrm{Mod}(X)$ as the category of quasi-coherent sheaves on X. Rosenberg in [**29**] has proved a reconstruction theorem which allows one to recover X from $\mathrm{Mod}(X)$ (generalizing work of Gabriel in the noetherian case).

Note however that the $\mathrm{Mod} : \mathrm{Sch} \to \mathrm{QSch}$ is not full. Indeed if X is scheme and $\mathcal{M} \in \mathrm{Mod}(X)$ then the functor $- \otimes_{\mathcal{O}_X} \mathcal{M}$ defines a morphism $\mathrm{Mod}(X) \to \mathrm{Mod}(X)$ in the category of quasi-schemes which in general will not correspond to a map $X \to X$.

In contrast Rosenberg has shown (unpublished) that the functor

$$\mathrm{Sch}/\operatorname{Spec}\mathbb{Z} \to \mathrm{Qsch}/\operatorname{Spec}\mathbb{Z}$$

which sends a scheme to its associated quasi-scheme is fully faithful in the sense of 2-categories. This result will not be used in the current paper. (For the definition of a relative two-category such as $\mathrm{Qsch}/\operatorname{Spec}\mathbb{Z}$ see Appendix A.)

Now let X be a quasi-scheme. We write $\mathrm{Alg}(X) = \mathrm{Alg}(\mathrm{Mod}(X))$, $\mathrm{Bimod}(X) = \mathrm{Bimod}(\mathrm{Mod}(X))$, etc .... Below by an *algebra on* X we will mean an object of $\mathrm{Alg}(X)$ unless otherwise specified. Likewise $\mathcal{A}$-$\mathcal{B}$-bimodules will in general be objects of $\mathrm{Bimod}(\mathcal{A}-\mathcal{B})$. By o_X we denote the identity functor on $\mathrm{Mod}(X)$ considered as an object of $\mathrm{Alg}(X)$. Obviously $\mathrm{Mod}(o_X) = \mathrm{Mod}(X)$, $\mathrm{Bimod}(o_X) = \mathrm{Bimod}(X)$.

If $\mathcal{A} \in \mathrm{Alg}(X)$ then we denote by $\operatorname{Spec}\mathcal{A}$ the object in Qsch/X given by the pair $(\mathrm{Mod}(\mathcal{A}), - \otimes_{o_X} \mathcal{A})$.

If X is a quasi-scheme then we define $\mathrm{Gralg}(X) = \mathrm{Gralg}(\mathrm{Mod}(X))$ and also $\mathrm{GRALG}(X) = \mathrm{GRALG}(\mathrm{Mod}(X))$. Below by a *graded algebra on* X we will mean an object of $\mathrm{Gralg}(X)$ unless otherwise specified. Likewise graded $\mathcal{A}$-$\mathcal{B}$-bimodules will in general be objects of $\mathrm{Bigr}(\mathcal{A} - \mathcal{B})$.

Related to the notion of relative categories is the notion of *enriched quasi-schemes*. An enriched quasi-scheme will be a pair $(X, \mathcal{O}_X)$ where X is a quasi-scheme and $\mathcal{O}_X$ is a some fixed object of $\mathrm{Mod}(X)$. A morphism $(Y, \mathcal{O}_Y) \to (X, \mathcal{O}_X)$ between enriched quasi-schemes is a pair (α, s) where $\alpha : Y \to X$ is a morphism between quasi-schemes and s is an isomorphism $s : \alpha^*(\mathcal{O}_X) \to \mathcal{O}_Y$. Note that if $(Y, \alpha) \in \mathrm{Qsch}/X$ and X is an enriched quasi-scheme then Y becomes canonically an enriched quasi-scheme if we put $\mathcal{O}_Y = \alpha^*\mathcal{O}_X$.

The prototype of an enriched quasi-scheme is $\operatorname{Spec} R = (\mathrm{Mod}(R), R_R)$ for a ring R. The following lemma will be used tacitly throughout the paper.

LEMMA 3.5.1. *Let R be a ring. Then the category* $\mathrm{Qsch}/\operatorname{Spec} R$ *is equivalent (as a two-category) to the category of R-linear enriched quasi-schemes. The equivalence is given by sending (Y, α) to $(Y, \alpha^*(R))$.*

If $(X, \mathcal{O}_X)$ is an enriched quasi-scheme and $\mathcal{U} \in \mathrm{Mod}(X)$ then we put $\Gamma(X, \mathcal{U}) = \mathrm{Hom}_{o_X}(\mathcal{O}_X, \mathcal{U})$. We say that $\mathcal{U}$ is generated by global sections if $\mathcal{U}$ is a quotient of some $\mathcal{O}_X^{\oplus I}$.

If $\mathcal{M}$ is a bimodule on X then we put $\mathcal{M}_{o_X} = \mathcal{O}_X \otimes_{o_X} \mathcal{M}$. We think of $\mathcal{M}_{o_X}$ as the "right structure" of $\mathcal{M}$. Care should be taken however since $\mathcal{M} \mapsto \mathcal{M}_{o_X}$ is apriori not an exact functor. This will not be a problem in our applications.

If $\mathcal{M}$ is a bimodule on X then we define $\Gamma(X, \mathcal{M}) = \Gamma(X, \mathcal{M}_{o_X})$. Again one should be careful since $\Gamma(X, -)$ is in general not left exact.

It is easy to see that if $\mathcal{A}$ is an algebra on X then $A = \Gamma(X, \mathcal{A})$ will be a ring. If $\mathcal{M}$ is an $\mathcal{A}$-module then $\Gamma(X, \mathcal{M})$ will be an A-module.

A quasi-scheme X will be called noetherian if $\mathrm{Mod}(X)$ is locally noetherian. For an enriched quasi-scheme, we also require that $\mathcal{O}_X$ is noetherian. A morphism $\alpha : X \to Y$ will be called noetherian if α^* preserves noetherian objects.

An algebra $\mathcal{A}$ on X is said to be noetherian if the functor $- \otimes_{o_X} \mathcal{A}$ preserves noetherian objects. Clearly if X is noetherian and $\mathcal{A}$ is noetherian then so is $\operatorname{Spec}\mathcal{A}$.

A this point we will introduce a convention that will be in force throughout this paper.

NOTATION . If $\mathcal{C} = \mathrm{Xyz}\cdots(\cdots)$ is a category then $\mathrm{xyz}\cdots(\cdots)$ stands for the full subcategory of $\mathcal{C}$ whose objects are the noetherian objects in $\mathcal{C}$.

3.6. Divisors

We will say that a morphism $\alpha : Y \to X$ is a closed embedding if α_* embeds $\mathrm{Mod}(Y)$ in $\mathrm{Mod}(X)$ as a closed subcategory.

Assume that X, Y are quasi-schemes where Y is embedded in X by a closed embedding. Just as in the commutative case, where one does not usually distinguishes between $\mathcal{O}_Y$ considered as a sheaf on X or on Y, we will write o_Y for both the identity functor on $\mathrm{Mod}(Y)$ and for the quotient of o_X associated to the inclusion $\mathrm{Mod}(Y) \subset \mathrm{Mod}(X)$.

For simplicity we assume that X is noetherian, although that is not strictly necessary. We denote

$$o_X(-Y) = \ker(o_X \to o_Y)$$

We say that Y is a divisor in X if $o_X(-Y)$ is invertible. In the rest of this section we will assume that Y is a divisor in X and we denote the inclusion morphism by i. If $\mathcal{M} \in \mathrm{Mod}(X)$ then we write $\mathcal{M}_Y$ for $\mathcal{M} \otimes_{o_X} o_Y$ and $\mathcal{M}(nY)$ for $\mathcal{M} \otimes_{o_X} o_X(nY)$, where $o_X(nY) = o_X(Y)^{\otimes n}$.

We denote the inclusion $o_X(-Y) \to o_X$ by t and we do the same with the induced maps $\mathcal{M}(-Y) \to \mathcal{M}$ for $\mathcal{M} \in \mathrm{Mod}(X)$. The normal bundle of Y in X is defined by $\mathcal{N}_{Y/X} = o_X(Y)/o_X$.

For use below we define a few categories.

$$\mathrm{tors}_Y(X) = \{\mathcal{M} \in \mathrm{mod}(X) \mid \text{There exist } n \text{ such that } t^n : \mathcal{M}(-nY) \to \mathcal{M} \text{ is zero}\}$$
$$\mathrm{iso}_Y(X) = \{\mathcal{M} \in \mathrm{mod}(X) \mid \text{The map } t : \mathcal{M}(-Y) \to \mathcal{M} \text{ is an isomorphism}\}$$

$\mathrm{Tors}_Y(X)$ and $\mathrm{Iso}_Y(X)$ will be the closures of $\mathrm{tors}_Y(X)$ and $\mathrm{iso}_Y(X)$ under direct unions.

LEMMA 3.6.1. $\mathrm{Tors}_Y(X)$ *and* $\mathrm{Iso}_Y(X)$ *are localizing subcategories of* $\mathrm{Mod}(X)$.

PROOF. Since $\mathrm{Mod}(X)$ is a locally noetherian category, it is easy to see that it is sufficient to show that $\mathrm{tors}_Y(X)$ and $\mathrm{iso}_Y(X)$ are Serre subcategories in $\mathrm{mod}(X)$. For $\mathrm{tors}_Y(X)$ this is clear, so we concentrate on $\mathrm{iso}_Y(X)$. It is clearly sufficient to show that $\mathrm{iso}_Y(X)$ is closed under taking subobjects. Let $\mathcal{M} \in \mathrm{iso}_Y(X)$ and let $\mathcal{N} \subset \mathcal{M}$. We have isomorphisms $t^n : \mathcal{M} \to \mathcal{M}(nY)$ and these yield an ascending chain of submodules in $\mathcal{M}$ given by $t^{-n}(\mathcal{N}(nY))$. Since $\mathcal{M}$ is noetherian this chain must stop. From this we easily obtain that $\mathcal{N} \in \mathrm{iso}_Y(X)$. □

We will say that the objects in $\mathrm{Tors}_Y(X)$ are *supported* on Y. If $\mathcal{M}(-Y) \to \mathcal{M}$ is monic then we will say that $\mathcal{M}$ is *Y-torsion free.*

REMARK 3.6.2. It is possible to define a notion of support for a general closed sub-quasi-scheme $Y \subset X$. However the current restricted notion is sufficient for our purposes.

We will use the following result.

PROPOSITION 3.6.3. *Let* $\mathcal{M} \in \mathrm{mod}(X)$. *Then the filtration*

$$\cdots \subset t^n(\mathcal{M}(-nY)) \subset t^{n-1}(\mathcal{M}(-(n-1)Y)) \subset \cdots \subset t(\mathcal{M}(-Y)) \subset \mathcal{M}$$

satisfies the Artin-Rees condition.

PROOF. Left to the reader. □

COROLLARY 3.6.4. *Let $\mathcal{M} \in \operatorname{mod}(X)$. Then $\mathcal{M}$ contains an Y-torsion free submodule $\mathcal{N}$, such that $\mathcal{M}/\mathcal{N}$ is supported on Y.*

PROOF. Let $\mathcal{T}$ be the maximal submodule of $\mathcal{M}$ supported on Y. Since $\mathcal{T}$ is noetherian we will have $t^n(\mathcal{T}(-nY)) = 0$ for some n. By Proposition 3.6.3 there will be some m such that $t^m(\mathcal{M}(-mY)) \cap \mathcal{T} \subset t^n(\mathcal{T}(-nY)) = 0$. Thus $\mathcal{N} = t^m(\mathcal{M}(-mY))$ is Y torsion free and $\mathcal{M}/\mathcal{N}$ is supported on Y. □

If $i : (Y, \mathcal{O}_Y) \to (X, \mathcal{O}_X)$ is a morphism of enriched quasi-schemes then we say that i makes Y into a divisor in X if the underlying map $Y \to X$ makes Y into a divisor in X in the sense of ordinary quasi-schemes and if in addition the induced map $\mathcal{O}_X(-Y) \to \mathcal{O}_X$ is monic.

3.7. Proj

Below X will be a noetherian quasi-scheme and $\mathcal{A} = \oplus_n \mathcal{A}_n \in \operatorname{Gralg}(X)$ will be noetherian. The definition of $\operatorname{Proj} \mathcal{A}$ is entirely similar to the ring case [**6**]. As before $\operatorname{Gr}(\mathcal{A})$ is the category of $\mathbb{Z}$-graded $\mathcal{A}$-modules, $\operatorname{Tors}(\mathcal{A})$ is the full subcategory of $\operatorname{Gr}(\mathcal{A})$ consisting of graded modules that are unions of right bounded modules and

$$\operatorname{Qgr}(\mathcal{A}) = \operatorname{Gr}(\mathcal{A}) / \operatorname{Tors}(\mathcal{A})$$

We use notation similar to that in [**6**]. Thus $\pi : \operatorname{Gr}(\mathcal{A}) \to \operatorname{Qgr}(\mathcal{A})$ is the quotient map. $\omega : \operatorname{Qgr}(\mathcal{A}) \to \operatorname{Gr}(\mathcal{A})$ is the right adjoint to π and $\tau : \operatorname{Gr}(\mathcal{A}) \to \operatorname{Tors}(\mathcal{A})$ is the functor which associates to every object its maximal torsion subobject. We put $\tilde{(-)} = \omega\pi$.

Let Pqsch/X ("polarized quasi-schemes") be the 2-category of triples (Y, α, s) where Y is a quasi-scheme, $\alpha : Y \to X$ a morphism and s an autoequivalence of $\operatorname{Mod}(Y)$. A morphism between triples $(Y, \alpha, s) \to (Z, \beta, t)$ is given by a morphism $\gamma : Y \to Z$, a natural isomorphism $\mu : \gamma^* \circ \beta^* \to \alpha^*$ and a natural isomorphism $\psi : s \circ \gamma^* \to \gamma^* \circ t$. We leave it to the reader to define natural isomorphisms between such triples. We now define $\operatorname{Proj} \mathcal{A}$ as the object of Pqsch/X given by the triple $(\operatorname{Qgr}(\mathcal{A}), \pi(- \otimes_{o_X} \mathcal{A}), s)$ where s is the shift functor on $\operatorname{Qgr}(\mathcal{A})$ obtained from the canonical shift on $\operatorname{Gr}(\mathcal{A})$. We will also denote by $\operatorname{Proj} \mathcal{A}$ the object in Qsch/X obtained by forgetting the shift.

If we denote by α the structure morphism $\operatorname{Proj} \mathcal{A} \to X$ then by definition, α^* is given by $\pi(- \otimes_{o_X} \mathcal{A})$. Since ω is the right adjoint to π we deduce that α_* is given by $\omega(-)_0$. In the sequel it will be convenient to use $\underline{\alpha}_*$ as a synonym for ω. Thus

$$\underline{\alpha}_*(\mathcal{M})_{o_X} = \oplus_n \alpha_*(\mathcal{M}(n))$$

Below we will generalize some of the results of [**6**] to our situation since we will need them. Usually we can simply copy the proofs in [**6**].

LEMMA 3.7.1. *Let $\mathcal{A}$, $\mathcal{B}$ be noetherian graded algebras on X, $\mathcal{N}$ a graded $\mathcal{A}$-module, $\mathcal{M}_1$, $\mathcal{M}_2$ graded $\mathcal{A} - \mathcal{B}$ bimodules.*

1. *Assume that $\mathcal{N}$ is right bounded and $(\mathcal{M}_1)_{o_X}$ is a quotient of $\mathcal{A} \otimes_{o_X} \mathcal{M}_1'$ where $\mathcal{M}_1'$ is a right bounded graded $o_X - o_X$-module. Then $\mathcal{N} \otimes_{\mathcal{A}} \mathcal{M}_1$ is right bounded.*
2. *Assume that $\mathcal{M}_1$ is right bounded. Then $\mathcal{N} \otimes_{\mathcal{A}} \mathcal{M}_1$ is torsion.*
3. *Assume that $\phi : \mathcal{M}_1 \to \mathcal{M}_2$ is a morphism of graded $\mathcal{A} - \mathcal{B}$ bimodule which is an isomorphism in high degree. Then* ker, coker$(\mathcal{N} \otimes_{\mathcal{A}} \mathcal{M}_1 \to \mathcal{N} \otimes_{\mathcal{A}} \mathcal{M}_2)$ *are torsion.*

PROOF. 1. This is trivial since $\mathcal{N} \otimes_{\mathcal{A}} \mathcal{M}_1$ is a quotient as graded o_X-bimodules of

$$\mathcal{N} \otimes_{\mathcal{A}} (\mathcal{A} \otimes_{o_X} \mathcal{M}_1') = \mathcal{N} \otimes_{o_X} \mathcal{M}_1'$$

2. This is a special case of (3).
3. Now write $\mathcal{N}$ as a quotient of modules of the form $\mathcal{P}_1 \to \mathcal{P}_0$ where $\mathcal{P}_{0,1}$ are direct sums of shifts of modules of the form $\mathcal{N}' \otimes_{o_X} \mathcal{A}$, $\mathcal{N}' \in \mathrm{Mod}(X)$ (located in degree zero). This is possible by Proposition 3.1.15(8).

   If $\mathcal{N} = \mathcal{N}' \otimes_{o_X} \mathcal{A}$ then the lemma is true because

$$(\mathcal{N}' \otimes_{o_X} \mathcal{A}) \otimes_{\mathcal{A}} \mathcal{M} = \mathcal{N}' \otimes_{o_X} {}_{o_X}\mathcal{M}$$

   for all $\mathcal{M} \in \mathrm{Bimod}(\mathcal{A} - \mathcal{B})$. Hence the map $\mathcal{N} \otimes_{\mathcal{A}} \mathcal{M}_1 \to \mathcal{N} \otimes_{\mathcal{A}} \mathcal{M}_2$ is an isomorphism in high degree. The general case follows from the fact that $- \otimes_{\mathcal{A}} \mathcal{M}_{1,2}$ is compatible with colimits. □

From this lemma one obtains the following corollary (cfr. [**6**, Prop. 2.5]).

COROLLARY 3.7.2. *Let $\phi : \mathcal{A} \to \mathcal{B}$ be a morphism of noetherian graded algebras on X such that ϕ is an isomorphism in high degree. Then the functors*

$$\mathrm{Gr}(\mathcal{A}) \to \mathrm{Gr}(\mathcal{B}) : \mathcal{N} \to \mathcal{N} \otimes_{\mathcal{A}} \mathcal{B} \tag{3.20}$$

and

$$\mathrm{Gr}(\mathcal{B}) \to \mathrm{Gr}(\mathcal{A}) : \mathcal{M} \to \mathcal{M}_{\mathcal{A}} \tag{3.21}$$

factor to give inverse equivalences between $\mathrm{Qgr}(\mathcal{A})$ *and* $\mathrm{Qgr}(\mathcal{B})$. *Furthermore the categories* $\mathrm{Proj}(\mathcal{A})$ *and* $\mathrm{Proj}(\mathcal{B})$ *are equivalent.*

PROOF. Assume that ϕ_n is an isomorphism for $n \geq n_0$. Then $\mathcal{B}$ is a quotient of $\mathcal{A} \oplus \mathcal{A} \otimes (\oplus_{n<n_0} \mathcal{B}_n(-n))$.

Assume that $\mathcal{M}$ is a $\mathcal{B}$-module, torsion as $\mathcal{A}$-module. Then $\mathcal{M}_{\mathcal{A}}$ is a quotient of $\oplus_{i \in I} \mathcal{M}_i$ where the $\mathcal{M}_i$ are right bounded $\mathcal{A}$-modules. Hence $\mathcal{M}$ is a quotient of $\oplus_{i \in I} \mathcal{M}_i \otimes_{\mathcal{A}} \mathcal{B}$. By lemma 3.7.1(1) all the $\mathcal{M}_i \otimes_{\mathcal{A}} \mathcal{B}$ are right bounded. Thus $\mathcal{M}$ is also torsion as $\mathcal{B}$-module.

Thus $\mathcal{B}$-torsion is equivalent to $\mathcal{A}$-torsion and so we will simply speak of torsion. (3.21) obviously preserves isomorphism mod torsion, so we concentrate on (3.20).

Assume that $\ker, \mathrm{coker}(\mathcal{M} \xrightarrow{\theta} \mathcal{N})$ are torsion for some $\mathcal{M}, \mathcal{N} \in \mathrm{Gr}(\mathcal{A})$. Then we have the following commutative diagram.

$$\begin{array}{ccc} \mathcal{M} & \longrightarrow & \mathcal{M} \otimes_{\mathcal{A}} \mathcal{B} \\ {\scriptstyle\theta}\downarrow & & \downarrow{\scriptstyle\theta\otimes 1} \\ \mathcal{N} & \longrightarrow & \mathcal{N} \otimes_{\mathcal{A}} \mathcal{B} \end{array}$$

According to lemma 3.7.1(3) the horizontal maps are isomorphisms modulo torsion. Since by hypotheses this is also true for θ, we obtain that $\theta \otimes 1$ is an isomorphism modulo torsion as well.

The reader now easily verifies that (3.20) and (3.21) are mutual inverses modulo torsion.

If $\mathcal{A}$, $\mathcal{B}$ are as in the previous proposition then (3.20) is clearly compatible with the structure morphisms and the shift functors on $\mathrm{Proj}\,\mathcal{A}$ and $\mathrm{Proj}\,\mathcal{B}$ whence $\mathrm{Proj}\,\mathcal{A}$ and $\mathrm{Proj}\,\mathcal{B}$ are equivalent. □

It follows from the previous proposition that one may restrict oneself to $\mathbb{N}$-graded algebras since $\operatorname{Proj} \mathcal{A} = \operatorname{Proj} \mathcal{A}_{\geq 0}$. In the sequel all graded algebras will be noetherian and $\mathbb{N}$-graded, unless otherwise specified. Note that just as in the ring case $\mathcal{A}$ noetherian implies $\mathcal{A}_{\geq 0}$ noetherian.

This allows us to use the following technically useful result which we state for further reference.

LEMMA 3.7.3. *Assume that $\mathcal{A}$ is noetherian and $\mathbb{N}$-graded. Then* $\operatorname{Tors}(A)$ *is closed under essential extensions and hence under injective hulls. Thus τ is* stable *in the sense of* [**31**]

Assume that $\mathcal{A}$ is $\mathbb{N}$-graded. Then if $\mathcal{M}, \mathcal{N} \in \operatorname{Gr}(\mathcal{A})$

$$\begin{aligned}\operatorname{Hom}_{\operatorname{Qgr}(\mathcal{A})}(\pi\mathcal{N}, \pi\mathcal{M}) &= \varinjlim_{\mathcal{N}/\mathcal{N}' \text{ torsion}} \operatorname{Hom}_{\operatorname{Gr}(\mathcal{A})}(\mathcal{N}', \mathcal{M}/\tau(\mathcal{M})) \\ &= \varinjlim_{\mathcal{N}/\mathcal{N}' \text{ torsion}} \operatorname{Hom}_{\operatorname{Gr}(\mathcal{A})}(\mathcal{N}', \mathcal{M})\end{aligned}$$

where the last equality follows from the fact that τ is stable (see [**31**]).

If $\mathcal{N} \in \operatorname{gr}(\mathcal{A})$ and if $\mathcal{N}/\mathcal{N}'$ is torsion then it is in fact right bounded, so $\mathcal{N}_{\geq n} \subset \mathcal{N}'$ for $n \geq 0$. So in that case

$$\operatorname{Hom}_{\operatorname{Qgr}(\mathcal{A})}(\pi\mathcal{N}, \pi\mathcal{M}) = \varinjlim_{n} \operatorname{Hom}_{\operatorname{Gr}(\mathcal{A})}(\mathcal{N}_{\geq n}, \mathcal{M}) \quad = \varinjlim \operatorname{Hom}_{\operatorname{Gr}(\mathcal{A})}(\mathcal{N}_{\geq n}, \mathcal{M}_{\geq n})$$

PROPOSITION 3.7.4. *Let $\mathcal{A} \in \operatorname{Gralg}(X)$ be noetherian and $\mathbb{N}$-graded. Then for* $\mathcal{M} \in \operatorname{Gr}(\mathcal{A})$

$$\tau(\mathcal{M}) = \varinjlim \underline{\mathcal{H}om}_{\mathcal{A}}(\mathcal{A}/\mathcal{A}_{\geq n}, \mathcal{M}) \tag{3.22}$$

$$\tilde{\mathcal{M}} = \varinjlim \underline{\mathcal{H}om}_{\mathcal{A}}(\mathcal{A}_{\geq n}, \mathcal{M}) \tag{3.23}$$

PROOF. By localization theory

$$\tau(\mathcal{M}) = \ker(\mathcal{M} \to \tilde{\mathcal{M}})$$

Now from the exact sequence of $\mathcal{A}$-bimodules

$$0 \to \mathcal{A}_{\geq n} \to \mathcal{A} \to \mathcal{A}/\mathcal{A}_{\geq n} \to 0$$

we obtain a left exact sequence

$$0 \to \underline{\mathcal{H}om}_{\mathcal{A}}(\mathcal{A}/\mathcal{A}_{\geq n}, \mathcal{M}) \to \underline{\mathcal{H}om}_{\mathcal{A}}(\mathcal{A}, \mathcal{M}) \to \underline{\mathcal{H}om}_{\mathcal{A}}(\mathcal{A}_{\geq n}, \mathcal{M})$$

Using the fact that $\underline{\mathcal{H}om}_{\mathcal{A}}(\mathcal{A}, \mathcal{M}) = \mathcal{M}$ and exactness of direct limits, it follows that it suffices to prove (3.23).

Recall that by definition $\tilde{\mathcal{M}} = \omega\pi\mathcal{M}$. So to prove (3.23) it is sufficient for every $\mathcal{N} \in \operatorname{Gr}(\mathcal{A})$ to construct a natural isomorphism between

$$\operatorname{Hom}_{\operatorname{Gr}(\mathcal{A})}(\mathcal{N}, \omega\pi\mathcal{M}) \tag{3.24}$$

and

$$\operatorname{Hom}_{\operatorname{Gr}(\mathcal{A})}(\mathcal{N}, \varinjlim \underline{\mathcal{H}om}_{\mathcal{A}}(\mathcal{A}_{\geq n}, \mathcal{M})) \tag{3.25}$$

and since $\operatorname{Gr}(\mathcal{A})$ is locally noetherian, it suffices to do this in fact for $\mathcal{N}$ noetherian. Now

$$(3.24) = \varinjlim_{n} \operatorname{Hom}_{\operatorname{Gr}(\mathcal{A})}(\mathcal{N}_{\geq n}, \mathcal{M}) \tag{3.26}$$

On the other hand, again because $\mathcal{N}$ is noetherian.

$$(3.25) = \varinjlim \operatorname{Hom}_{\operatorname{Gr}(\mathcal{A})}(\mathcal{N} \otimes_{\mathcal{A}} \mathcal{A}_{\geq n}, \mathcal{M}) \tag{3.27}$$

We have to make (3.26) isomorphic to (3.27). Now $\mathcal{N}$ is noetherian and hence left bounded, so by replacing $\mathcal{N}$ by some shift (and doing the same with $\mathcal{M}$) we may assume that $\mathcal{N}$ is in fact $\mathbb{N}$-graded. Using (3.2) and lemma 3.1.11 it is now sufficient to show that

$$\begin{aligned} \mathcal{K}_n &= \ker(\mathcal{N} \otimes_{\mathcal{A}} \mathcal{A}_{\geq n} \to \mathcal{N}_{\geq n}) \\ \mathcal{C}_n &= \operatorname{coker}(\mathcal{N} \otimes_{\mathcal{A}} \mathcal{A}_{\geq n} \to \mathcal{N}_{\geq n}) \end{aligned}$$

are torsion inverse systems (see Definition 3.1.9).

Now we write $\mathcal{N}$ as quotient of $\mathcal{P}_1 \to \mathcal{P}_0$ where the $\mathcal{P}_i$ are finite direct sums of negative shifts of modules of the form $\mathcal{N}' \otimes_{o_X} \mathcal{A}$. Working in the abelian category of inverse systems modulo torsion it is sufficient to prove that $\mathcal{K}_n$ and $\mathcal{C}_n$ are torsion inverse systems in the case $\mathcal{N} = \mathcal{N}' \otimes_{o_X} \mathcal{A}(-m)$. In that case

$$\begin{aligned} \mathcal{K}_n &= \ker(\mathcal{N}' \otimes_{o_X} \mathcal{A}_{\geq n} \to \mathcal{N}' \otimes_{o_X} \mathcal{A}_{\geq n-m})(-m) \\ \mathcal{C}_n &= \operatorname{coker}(\mathcal{N}' \otimes_{o_X} \mathcal{A}_{\geq n} \to \mathcal{N}' \otimes_{o_X} \mathcal{A}_{\geq n-m})(-m) \end{aligned}$$

It is now clear that in this case actually $\mathcal{K}_n = 0$ and $\mathcal{C}_n$ lives in degrees $[n, n+m-1]$. So $(\mathcal{C}_n)_n$ is a torsion inverse system as well. □

3.8. Condition "χ" and cohomological dimension

Below X will be a noetherian quasi-scheme and $\mathcal{A}$, $\mathcal{B}$ will be $\mathbb{N}$-graded noetherian objects in $\operatorname{Gralg}(X)$. $\alpha : \operatorname{Proj} \mathcal{A} \to X$, $\beta : \operatorname{Proj} \mathcal{B} \to X$ will be the structure morphisms. π, ω, τ refer to $\mathcal{A}$. For $\mathcal{B}$ we use notations such as $\pi_{\mathcal{B}}$, $\omega_{\mathcal{B}}$, $\tau_{\mathcal{B}}$.

The following is clear.

LEMMA 3.8.1. *If* $\mathcal{M} \in \operatorname{Gr}(\mathcal{A})$ *then* $R^i\tau\mathcal{M} \in \operatorname{Tors}(\mathcal{A})$.

In order to develop a non-commutative version of projective geometry, the following definition was proposed in [**6**].

DEFINITION 3.8.2. $\mathcal{A}$ satisfies χ if for every $\mathcal{M} \in \operatorname{gr}(\mathcal{A})$ one has that $(R^i\tau\mathcal{M})_{\geq 0}$ is contained in $\operatorname{tors}(\mathcal{A})$ for all $i \geq 0$.

If $\mathcal{M} \in \operatorname{Gr}(\mathcal{A})$ then we have a triangle

$$R\tau\mathcal{M} \to \mathcal{M} \to R\omega(\pi\mathcal{M}) \xrightarrow{[1]}$$

from which one deduces the exact sequence

$$0 \to \tau\mathcal{M} \to \mathcal{M} \to \tilde{\mathcal{M}} \to R^1\tau\mathcal{M} \to 0 \tag{3.28}$$

and isomorphisms

$$R^{i+1}\tau\mathcal{M} = R^i\omega(\pi\mathcal{M})), \qquad i \geq 1 \tag{3.29}$$

So we obtain :

LEMMA 3.8.3. *$\mathcal{A}$ satisfies χ if and only if for all $\mathcal{M} \in \operatorname{gr}\mathcal{A}$ one has*

1. $R^i\omega(\pi\mathcal{M})_{\geq 0} \in \operatorname{tors}(\mathcal{A})$, $i \geq 1$.
2. $(\tilde{\mathcal{M}}/\mathcal{M})_{\geq 0} \in \operatorname{tors}(\mathcal{A})$

One should view condition χ as a kind of ampleness condition. This becomes clearer if one makes the following definitions.

DEFINITION 3.8.4. A morphism $\alpha : Y \to X$ of noetherian quasi-schemes is *proper* if $R^i\alpha_*$ sends $\mathrm{mod}(Y)$ to $\mathrm{mod}(X)$ for all i.

DEFINITION 3.8.5. Let (Y, α, s) be an object in Pqsch/X. Assume that Y is noetherian. Then we say that s is *relatively ample* if the following two conditions hold for all $\mathcal{M} \in \mathrm{mod}(Y)$.

1. For $i > 0$ one has
$$R^i\alpha_*(s^n\mathcal{M}) = 0$$
for $n \gg 0$.
2. The adjoint map
$$\alpha^*\alpha_*s^n\mathcal{M} \to s^n\mathcal{M}$$
is epic for $n \gg 0$.

REMARK 3.8.6. This definition of ampleness is much more restrictive than the one used in [**6**].

We say that $\mathcal{A}$ is generated in degree one if the multiplication map $\mathcal{A}_m \otimes_{o_X} \mathcal{A}_n \to \mathcal{A}_{m+n}$ is epic for all $m, n \geq 0$.

The following is a consequence of lemma 3.8.3.

PROPOSITION 3.8.7. *Let $(Y, \alpha, s) = \mathrm{Proj}\,\mathcal{A}$. If $\mathcal{A}$ satisfies χ then α is proper and if $\mathcal{A}$ is in addition generated in degree one then s is relatively ample.*

In the following we will often need the following technical condition on an object $\mathcal{M} \in \mathrm{BIGR}(\mathcal{A} - o_X)$:

(fin) $\mathcal{M}$ has a presentation $\mathcal{P}_1 \to \mathcal{P}_0$ where the $\mathcal{P}_i$ are finite sums of shifts of $\mathcal{A} - o_X$-bimodules of the form $\mathcal{A} \otimes_{o_X} \mathcal{P}'$ where the $\mathcal{P}'$ are coherent objects (see 3.1) in $\mathrm{BIMOD}(o_X - o_X)$ (located in degree zero).

We should think of (fin) as a kind of "finite presentation" condition. (fin) is useful because of the following lemma

LEMMA 3.8.8. *Assume that $\mathcal{M} \in \mathrm{BIGR}(\mathcal{A} - o_X)$ satisfies (fin) and $\mathcal{N}$ is an object in $\mathrm{BIGR}(\mathcal{B} - \mathcal{A})$ or in $\mathrm{Gr}(\mathcal{A})$. Assume furthermore that $\mathcal{M}$ is $\mathbb{N}$-graded. Then the kernel and the cokernel of*
$$\mathcal{N}_{\geq n} \otimes_{\mathcal{A}} \mathcal{M} \to (\mathcal{N} \otimes_{\mathcal{A}} \mathcal{M})_{\geq n}$$
are torsion inverse systems (cfr. Definition 3.1.9).

Here is another obvious application.

LEMMA 3.8.9. *Let $f : \mathcal{A} \to \mathcal{B}$ be a morphism in $\mathrm{Gralg}(X)$ and assume that $\mathcal{B}$ satisfies (fin). Let $\mathcal{M} \in \mathrm{Gr}(\mathcal{B})$. Then $\tau_{\mathcal{A}}(\mathcal{M}_{\mathcal{A}}) = (\tau_{\mathcal{B}}(\mathcal{M}))_{\mathcal{A}}$.*

PROOF. Using Proposition 3.7.4 and adjointness this amounts to showing that the inverse systems $\mathcal{A}/\mathcal{A}_{\geq n} \otimes_{\mathcal{A}} \mathcal{B}$ and $\mathcal{B}/\mathcal{B}_{\geq n}$ are equivalent modulo torsion inverse systems. Since $\mathcal{A}/\mathcal{A}_{\geq n} \otimes_{\mathcal{A}} \mathcal{B} = \mathrm{coker}(\mathcal{A}_{\geq n} \otimes_{\mathcal{A}} \mathcal{B} \to \mathcal{B})$ this follows directly from lemma 3.8.8. □

In the sequel we will need a result like (3.30). The hypotheses under which we can prove this are unfortunately quite technical and almost certainly too restrictive.

PROPOSITION 3.8.10. *Let $f : \mathcal{A} \to \mathcal{B}$ be a morphism in* $\mathrm{Gralg}(X)$*. Assume that $\mathcal{B}$ satisfies (fin) and furthermore that there are graded flat $\mathcal{A} - o_X$-bimodules $(\mathcal{F}_i)_i$ satisfying (fin), together with a long exact sequence*

$$\cdots \to (\mathcal{F}_2)_{\geq p} \to (\mathcal{F}_1)_{\geq p} \to (\mathcal{F}_0)_{\geq p} \to ({}_{\mathcal{A}}\mathcal{B}_{o_X})_{\geq p} \to 0$$

for a certain $p \in \mathbb{Z}$. Then for $\mathcal{M} \in \mathrm{Gr}(\mathcal{B})$ one has

$$R^i\tau_{\mathcal{A}}(\mathcal{M}_{\mathcal{A}}) = (R^i\tau_{\mathcal{B}}(\mathcal{M}))_{\mathcal{A}} \tag{3.30}$$

PROOF. Both sides of (3.30) are δ-functors. Hence is suffices to show that the two sides of (3.30) are zero for $i > 0$ and naturally isomorphic for $i = 0$, when evaluated on injectives.

Actually it is not necessary to take all injectives, but only a cogenerating set. Thus we take for our injectives all modules of the form

$$F = \underline{\mathcal{H}om}_{\mathcal{A}}({}_{\mathcal{B}}\mathcal{B}_{\mathcal{A}}, E)$$

where $E \in \mathrm{Inj}(\mathcal{A})$. Clearly by Proposition 3.7.4, lemma 3.8.9 and adjointness

$$\mathrm{RHS}(3.30) = \begin{cases} \varinjlim \underline{\mathcal{H}om}_{\mathcal{A}}(\mathcal{A}/\mathcal{A}_{\leq n} \otimes_{\mathcal{A}} {}_{\mathcal{A}}\mathcal{B}_{\mathcal{A}}, E) & \text{if } i = 0 \\ 0 & \text{otherwise} \end{cases}$$

On the other hand by the definition of $\underline{\mathcal{T}or}$

$$\begin{aligned} \mathrm{LHS}(3.30) &= \varinjlim \underline{\mathcal{E}xt}^i_{\mathcal{A}}(\mathcal{A}/\mathcal{A}_{\geq n}, F_{\mathcal{A}}) \\ &= \varinjlim \underline{\mathcal{E}xt}^i_{\mathcal{A}}(\mathcal{A}/\mathcal{A}_{\geq n}, \underline{\mathcal{H}om}_{\mathcal{A}}({}_{\mathcal{A}}\mathcal{B}_{\mathcal{A}}, E)) \\ &= \varinjlim \underline{\mathcal{H}om}_{\mathcal{A}}(\underline{\mathcal{T}or}^{\mathcal{A}}_i(\mathcal{A}/\mathcal{A}_{\geq n}, {}_{\mathcal{A}}\mathcal{B}_{\mathcal{A}}), E) \end{aligned}$$

So the two sides of (3.30) certainly agree for $i = 0$. In general we have to show for $i > 0$ that

$$\underline{\mathcal{T}or}^{\mathcal{A}}_i(\mathcal{A}/\mathcal{A}_{\geq n}, {}_{\mathcal{A}}\mathcal{B}_{\mathcal{A}}) \tag{3.31}$$

is a torsion inverse system.

From the fact that the $\mathcal{F}_i$ are flat we can deduce that, up to inverse systems of the form

$$\underline{\mathcal{T}or}^{\mathcal{A}}_i(\mathcal{A}/\mathcal{A}_{\geq n}, \mathcal{T}) \tag{3.32}$$

with $\mathcal{T}_{o_X}$ coherent, (3.31) is given by the middle homology of

$$\mathcal{A}/\mathcal{A}_{\geq n} \otimes_{\mathcal{A}} (\mathcal{F}_{i+1})_{\geq p} \to \mathcal{A}/\mathcal{A}_{\geq n} \otimes_{\mathcal{A}} (\mathcal{F}_i)_{\geq p} \to \mathcal{A}/\mathcal{A}_{\geq n} \otimes_{\mathcal{A}} (\mathcal{F}_{i-1})_{\geq p} \tag{3.33}$$

By shifting if necessary we may assume that $\mathcal{F}_{i+1}$, $\mathcal{F}_i$, $\mathcal{F}_{i-1}$ are $\mathbb{N}$-graded.

Now assume in general that $\mathcal{F}$ is an $\mathbb{N}$-graded object in $\mathrm{BIGR}(\mathcal{A} - o_X)$, satisfying (fin). Then

$$\mathcal{A}/\mathcal{A}_{\geq n} \otimes_{\mathcal{A}} \mathcal{F}_{\geq p} = \mathrm{coker}(\mathcal{A}_{\geq n} \otimes_{\mathcal{A}} \mathcal{F}_{\geq p} \to \mathcal{F}_{\geq p})$$

Now up to inverse systems of the form

$$\underline{\mathcal{T}or}^{\mathcal{A}}_i(\mathcal{A}_{\geq n}, \mathcal{F}/\mathcal{F}_{\geq p}) \tag{3.34}$$

we have

$$\mathcal{A}_{\geq n} \otimes_{\mathcal{A}} \mathcal{F}_{\geq p} = \mathcal{A}_{\geq n} \otimes_{\mathcal{A}} \mathcal{F}$$

Then using lemma 3.8.8 we see that $\mathcal{A}_{\geq n} \otimes_{\mathcal{A}} \mathcal{F}$ is, modulo torsion inverse systems, equal to $\mathcal{F}_{\geq n}$. Assembling everything we find that up to torsion inverse systems and inverse systems of the form (3.32) and (3.34) $\mathcal{A}/\mathcal{A}_{\geq n} \otimes_{\mathcal{A}} \mathcal{F}_{\geq p}$ is equal to $\mathcal{F}_{\geq p}/\mathcal{F}_{\geq n}$ (where we let the inverse systems start with $n = p$).

So the middle homology of (3.33) is, up to torsion inverse systems and inverse systems of the form (3.32) and (3.34), equal to the middle homology of

$$\mathcal{F}_{i+1}/(\mathcal{F}_{i+1})_{\geq n} \to \mathcal{F}_i/(\mathcal{F}_i)_{\geq n} \to \mathcal{F}_{i-1}/(\mathcal{F}_{i-1})_{\geq n}$$

Since this is in fact an exact complex we are left with showing that (3.32) and (3.34) are torsion.

Let us for example consider (3.32). It is easy to verify directly from the definitions that

$$(\text{“}\varprojlim\text{”}\,\underline{\mathcal{T}or}_i^{\mathcal{A}}(\mathcal{A}/\mathcal{A}_{\geq n}, \mathcal{T}))_{o_X} = \text{“}\varprojlim\text{”}\,\underline{\mathcal{T}or}_i^{\mathcal{A}}(\mathcal{A}/\mathcal{A}_{\geq n}, \mathcal{T}_{o_X})$$

Now as in the proof of lemma 3.1.18 one verifies that $\underline{\mathcal{T}or}_i^{\mathcal{A}}(\mathcal{A}/\mathcal{A}_{\geq n}, \mathcal{T}_{o_X})$ is coherent as weak $\mathcal{A} - o_X$-bimodule. Hence $\underline{\mathcal{H}om}_{o_X}(\underline{\mathcal{T}or}_i^{\mathcal{A}}(\mathcal{A}/\mathcal{A}_{\geq n}, \mathcal{T}_{o_X}), -)$ commutes with direct sums. Thus it suffices to show that

$$\varinjlim \underline{\mathcal{H}om}_{o_X}(\underline{\mathcal{T}or}_i^{\mathcal{A}}(\mathcal{A}/\mathcal{A}_{\geq n}, \mathcal{T}_{o_X}), E) = 0 \tag{3.35}$$

for E an injective object in $\mathrm{Mod}(X)$, viewed as a graded object concentrated in a single degree. Now (3.35) becomes equal to

$$\varinjlim \underline{\mathcal{E}xt}_{\mathcal{A}}^i(\mathcal{A}/\mathcal{A}_{\geq n}, \underline{\mathcal{H}om}_{o_X}(\mathcal{T}_{o_X}, E)) \tag{3.36}$$

Clearly $\underline{\mathcal{H}om}_{o_X}(\mathcal{T}_{o_X}, E)$ is right bounded and hence torsion. Now the fact that $\mathrm{Tors}(\mathcal{A})$ is stable (lemma 3.7.3) together with Proposition 3.7.4 easily implies that (3.36) is zero. □

Now we translate (3.30) into more geometrical language. First we indicate when a map $f : \mathcal{A} \to \mathcal{B}$ defines a dual map $\bar{f} : \mathrm{Proj}(\mathcal{B}) \to \mathrm{Proj}(\mathcal{A})$

PROPOSITION 3.8.11. *Let $f : \mathcal{A} \to \mathcal{B}$ be a morphism in* Gralg(X) *such that ${}_{\mathcal{A}}\mathcal{B}_{o_X}$ satisfies (fin). Then the functors*

$$- \otimes_{\mathcal{A}} \mathcal{B} : \mathrm{Gr}(\mathcal{A}) \to \mathrm{Gr}(\mathcal{B}) \tag{3.37}$$

$$(-)_{\mathcal{A}} : \mathrm{Gr}(\mathcal{B}) \to \mathrm{Gr}(\mathcal{A}) \tag{3.38}$$

*factor through "*Qgr*" and in this way define respectively $\bar{f}^*$ and $\bar{f}_*$ for a morphism $\bar{f} : \mathrm{Proj}(\mathcal{B}) \to \mathrm{Proj}(\mathcal{A})$. In this case $\bar{f}_*$ is exact.*

PROOF. That (3.38) factors through "Qgr" as well as the exactness of $\bar{f}_*$ is clear so we concentrate on (3.37). We have to show that if a map $\mathcal{M} \to \mathcal{N}$ in $\mathrm{Mod}(\mathcal{A})$ has torsion kernel and cokernel then the same is true for

$$\mathcal{M} \otimes_{\mathcal{A}} \mathcal{B} \to \mathcal{N} \otimes_{\mathcal{A}} \mathcal{B}$$

Now $\mathrm{Mod}(\mathcal{A})$ is locally noetherian and $- \otimes_{\mathcal{A}} \mathcal{B}$ is compatible with direct limits, so we may restrict ourselves to the case where $\mathcal{M}$, $\mathcal{N}$ are noetherian and hence $\mathcal{M}_n \to \mathcal{N}_n$ is an isomorphism for $n \gg 0$.

Now using the fact that $\mathcal{B}$ satisfies (fin) it suffices to show that

$$\mathcal{M} \otimes_{\mathcal{A}} (\mathcal{A} \otimes_{o_X} \mathcal{B}') \to \mathcal{N} \otimes_{\mathcal{A}} (\mathcal{A} \otimes_{o_X} \mathcal{B}') \tag{3.39}$$

has torsion kernel and cokernel for all $\mathcal{B}' \in \mathrm{BIMOD}(o_X - o_X)$. However since $- \otimes_{\mathcal{A}} (\mathcal{A} \otimes_{o_X} \mathcal{B}') = - \otimes_{o_X} \mathcal{B}'$ it is clear that (3.39) is an isomorphism in high degree.

It remains to verify the adjointness of (3.37) and (3.38) when defined on "QGr". To this end we have to construct for $\mathcal{M} \in \mathrm{Gr}(\mathcal{A})$ and $\mathcal{N} \in \mathrm{Gr}(\mathcal{B})$ a natural isomorphism

$$\mathrm{Hom}_{\mathrm{Qgr}(\mathcal{B})}(\pi(\mathcal{M} \otimes_{\mathcal{A}} \mathcal{B}), \pi\mathcal{N}) \cong \mathrm{Hom}_{\mathrm{Qgr}(\mathcal{A})}(\pi\mathcal{M}, \pi(\mathcal{N}_{\mathcal{A}}))$$

and as usual it suffices to do this for $\mathcal{M}$ noetherian. Then

$$\begin{aligned}\mathrm{Hom}_{\mathrm{Qgr}(\mathcal{B})}(\pi(\mathcal{M} \otimes_{\mathcal{A}} \mathcal{B}), \pi\mathcal{N}) &= \varinjlim \mathrm{Hom}_{\mathrm{Gr}(\mathcal{B})}((\mathcal{M} \otimes_{\mathcal{A}} \mathcal{B})_{\geq n}, \mathcal{N}) \\ &= \varinjlim \mathrm{Hom}_{\mathrm{Gr}(\mathcal{B})}(\mathcal{M}_{\geq n} \otimes_{\mathcal{A}} \mathcal{B}, \mathcal{N}) \\ &= \varinjlim \mathrm{Hom}_{\mathrm{Gr}(\mathcal{A})}(\mathcal{M}_{\geq n}, \mathcal{N}_{\mathcal{A}}) \\ &= \mathrm{Hom}_{\mathrm{Qgr}(\mathcal{A})}(\pi\mathcal{M}, \pi(\mathcal{N}_{\mathcal{A}}))\end{aligned}$$

The second equality follows from the fact that $\mathcal{B}$ satisfies (fin) as $\mathcal{A} - o_X$-bimodule and thus we can apply lemma 3.8.8 □

Thus we obtain

PROPOSITION 3.8.12. *Let $f : \mathcal{A} \to \mathcal{B}$ be a morphism in* $\mathrm{Gralg}(X)$ *and assume that $\mathcal{B}$ is as in Proposition 3.8.10. Let $\bar{f} : \mathrm{Proj}\,\mathcal{B} \to \mathrm{Proj}\,\mathcal{A}$ be defined by $\bar{f}^*(\pi\mathcal{M}) = \pi(\mathcal{M} \otimes_{\mathcal{A}} \mathcal{B})$ for $\mathcal{M} \in \mathrm{Gr}(\mathcal{A})$. Then for $\mathcal{N} \in \mathrm{Qgr}(\mathcal{B})$*

$$R^i\beta_*(\mathcal{N}) = R^i\alpha_*(\bar{f}_*\mathcal{N})$$

PROOF. Assume $\mathcal{N} = \pi\mathcal{N}'$, $\mathcal{N}' \in \mathrm{Gr}(\mathcal{B})$. Summing over shifts, it is sufficient to show

$$R^i\omega_{\mathcal{B}}(\pi\mathcal{N}')_{\mathcal{A}} = R^i\omega_{\mathcal{A}}(\pi(\mathcal{N}'_{\mathcal{A}}))$$

If $i \geq 1$ then by (3.29) this is equivalent to

$$R^{i+1}\tau_{\mathcal{B}}(\mathcal{N}')_{\mathcal{A}} = R^{i+1}\tau_{\mathcal{A}}(\mathcal{N}'_{\mathcal{A}})$$

which we have shown. The case $i = 0$ follows from considering (3.28) and the 5-lemma. □

Now we use Proposition 3.8.10 to prove the following result

PROPOSITION 3.8.13. *Assume that we have a map $f : \mathcal{A} \to \mathcal{B}$ of $\mathbb{N}$-graded algebras on X. Assume furthermore that there is an exact sequence*

$$0 \to I \to \mathcal{A} \to {}_{\mathcal{A}}\mathcal{B}_{\mathcal{A}} \to 0$$

in $\mathrm{Bigr}(\mathcal{A} - \mathcal{A})$ *with I a graded invertible $\mathcal{A}$-bimodule, satisfying (fin) and living in degree ≥ 1. Assume in addition that I^{-1} satisfies (fin). Then*

1. *If $\mathcal{B}$ is noetherian then so is $\mathcal{A}$.*

Assume now that $\mathcal{B}$ is noetherian.

2. *If $\mathcal{B}$ satisfies χ then so does $\mathcal{A}$.*
3. *Let "cd" stand for cohomological dimension. Then*

$$\mathrm{cd}\,\tau_{\mathcal{B}} + 1 \geq \mathrm{cd}\,\tau_{\mathcal{A}} \geq \mathrm{cd}\,\tau_{\mathcal{B}}$$

PROOF. 1. We have to show that if $\mathcal{M} \in \mathrm{Mod}(X)$ is noetherian then so is $\mathcal{M} \otimes_{o_X} \mathcal{A}$.

Now if $\mathcal{N} \in \mathrm{Gr}(\mathcal{A})$ is left bounded then a variation of the classical argument by Hilbert shows that if $\mathcal{N} \otimes_{\mathcal{A}} \mathcal{B}$ is noetherian then the same holds for $\mathcal{N}$. Since by hypotheses

$$(\mathcal{M} \otimes_{o_X} \mathcal{A}) \otimes_{\mathcal{A}} \mathcal{B} = \mathcal{M} \otimes_{o_X} \mathcal{B}$$

is noetherian we are done.

2. We start by observing that $\mathcal{B}$ clearly satisfies the hypotheses of Proposition 3.8.10, so we can use that proposition.

In the sequel we will denote by G the functor $-\otimes_{\mathcal{A}} I$ and t the natural transformation $G \to \mathrm{id}_{\mathrm{Gr}(\mathcal{A})}$ coming from the inclusion $I \to \mathcal{A}$. Furthermore $G^n(\mathcal{N}) \xrightarrow{t^n} \mathcal{N}$ is by definition obtained from tensoring $I^n \xrightarrow{t^n} \mathcal{A}$ with $\mathcal{N}$. By $\mathcal{N}t^n$ we denote the image of this map. We say that $\mathcal{N}$ is annihilated by t^n if $\mathcal{N}t^n = 0$. $\mathcal{N}$ is t-torsion if it is the union of subobjects annihilated by some t^n. Similarly $\mathcal{N}$ is t-torsion free if multiplication by t is monic.

It is easy to see that the functors

$$-\otimes_{\mathcal{A}} \mathcal{B} : \mathrm{Gr}(\mathcal{A}) \to \mathrm{Gr}(\mathcal{B})$$
$$(-)_{\mathcal{A}} : \mathrm{Gr}(\mathcal{B}) \to \mathrm{Gr}(\mathcal{A})$$

induce inverse equivalences between $\mathrm{Gr}(\mathcal{B})$ and the full subcategory of $\mathrm{Gr}(\mathcal{A})$ consisting of objects annihilated by t. So we will identify these two categories.

Since I satifies (fin) it is easy to see that $G(\mathrm{Tors}(\mathcal{A})) \subset \mathrm{Tors}(\mathcal{A})$. Since I^{-1} satisfies (fin) as well we obtain $G(\mathrm{Tors}(\mathcal{A})) = \mathrm{Tors}(\mathcal{A})$. In particular G commutes with $\tau_{\mathcal{A}}$. Since G is an autoequivalence this easily yields that G commutes with $R^i\tau_{\mathcal{A}}$. This fact is used in the computation below.

Let $\mathcal{M} \in \mathrm{gr}(\mathcal{A})$. As usual $\mathcal{M}$ is an extension

$$0 \to \mathcal{M}_1 \to \mathcal{M} \to \mathcal{M}_2 \to 0$$

where $\mathcal{M}_1$ is t-torsion and $\mathcal{M}_2$ is t-torsion free. Now $\mathcal{M}$ is noetherian and hence so is $\mathcal{M}_1$. Thus $\mathcal{M}_1 t^n = 0$ for some n and in particular we can write $\mathcal{M}_1$ as an extension of objects annihilated by t.

We conclude that to verify condition χ we have to show that $R^i\tau_{\mathcal{A}}(\mathcal{N})_{\geq 0}$ is right bounded and noetherian for two classes of noetherian graded $\mathcal{A}$-modules :

(a) Those that are annihilated by t.

(b) Those that are t-torsion free.

Let us first treat (a). If $\mathcal{N}t = 0$ then by the previous discussion $\mathcal{N} = \mathcal{N}'_{\mathcal{A}}$ for some $\mathcal{N}' \in \mathrm{gr}(\mathcal{B})$. Thus

$$\begin{aligned} R^i\tau_{\mathcal{A}}(\mathcal{N}) &= R^i\tau_{\mathcal{A}}(\mathcal{N}'_{\mathcal{A}}) \\ &= R^i\tau_{\mathcal{B}}(\mathcal{N}')_{\mathcal{A}} \qquad \text{(Prop. 3.8.10)} \end{aligned}$$

which shows what we want.

Now consider the case that $\mathcal{N}$ is torsion free. We have an exact sequence in $\mathrm{Gr}(\mathcal{A})$

$$0 \to G(\mathcal{N}) \xrightarrow{t} \mathcal{N} \to \mathcal{N}/\mathcal{N}t \to 0$$

and again $\mathcal{N}/\mathcal{N}t = \mathcal{N}''_{\mathcal{A}}$ for some $\mathcal{N}'' \in \mathrm{gr}(\mathcal{B})$. This gives us the following exact sequence (using Prop. 3.8.10)

$$R^{i-1}\tau_{\mathcal{B}}(\mathcal{N}'')_{\mathcal{A}} \to G(R^i\tau_{\mathcal{A}}(\mathcal{N})) \xrightarrow{t} R^i\tau_{\mathcal{A}}(\mathcal{N}) \to R^i\tau_{\mathcal{B}}(\mathcal{N}'')_{\mathcal{A}} \tag{3.40}$$

Lemma 3.8.1 implies $R^i\tau_{\mathcal{A}}(\mathcal{N}) \in \mathrm{Tors}(\mathcal{A})$. $R^i\tau_{\mathcal{A}}(\mathcal{N})$ is t-torsion since I lives purely in positive degree. On the other hand we obtain from (3.40) and the fact that $\mathcal{B}$ satisfies χ that $R^i\tau_{\mathcal{A}}(\mathcal{N})_{\geq n}$ is t-torsion free for $n \gg 0$.

Combining this we obtain that $R^i\tau_{\mathcal{A}}(\mathcal{N})_{\geq n} = 0$ for $n \gg 0$ which shows that $R^i\tau_{\mathcal{A}}(\mathcal{N})$ is right bounded.

From (3.40) we obtain exact sequences

$$0 \to \text{noetherian} \to G(R^i\tau_{\mathcal{A}}(\mathcal{N}))_{\geq m} \xrightarrow{t} R^i\tau_{\mathcal{A}}(\mathcal{N})_{\geq m}$$

Since I lives in degrees ≥ 1 we have $G(R^i\tau_{\mathcal{A}}(\mathcal{N})_{\geq m-1}) \subset G(R^i\tau_{\mathcal{A}}(\mathcal{N}))_{\geq m}$ and furthermore since subobjects of noetherian objects are noetherian we obtain an exact sequence

$$0 \to \text{noetherian} \to G(R^i\tau_{\mathcal{A}}(\mathcal{N})_{\geq m-1}) \xrightarrow{t} R^i\tau_{\mathcal{A}}(\mathcal{N})_{\geq m}$$

By descending induction on m we find that $R^i\tau_{\mathcal{A}}(\mathcal{N})_{\geq 0}$ is noetherian.

3. From Proposition 3.8.10 it follows immediately that $\operatorname{cd}\tau_{\mathcal{A}} \geq \operatorname{cd}\tau_{\mathcal{B}}$. We therefore concentrate on the other inequality. Assume $\mathcal{N} \in \operatorname{Gr}(\mathcal{A})$, $p = \operatorname{cd}\tau_{\mathcal{B}}$. We have to show $R^q\tau_{\mathcal{A}}(\mathcal{N}) = 0$, $q > p+1$.

   Since $\tau_{\mathcal{A}}$ commutes with direct limits and $\operatorname{Gr}(\mathcal{A})$ is locally noetherian, $R^q\tau_{\mathcal{A}}$ also commutes with direct limits [**17**]. Hence we may assume that $\mathcal{N}$ is noetherian. Using the same reduction as in (2) we may assume that either $\mathcal{N}t = 0$ or $\mathcal{N}$ is t-torsion free, The first case is trivial by Proposition 3.8.10 so we look at the last case. Using (3.40) for $i = q$ we find that $G(R^q\tau_{\mathcal{A}}(\mathcal{N})) \to R^q\tau_{\mathcal{A}}(\mathcal{N})$ is an isomorphism. Since on the other hand $R^q\tau_{\mathcal{A}}(\mathcal{N})$ is t-torsion we conclude $R^q\tau_{\mathcal{A}}(\mathcal{N}) = 0$. □

Proposition 3.8.13(3) can be stated in more geometric terms.

COROLLARY 3.8.14. *Assume that $\mathcal{A}$, $\mathcal{B}$ are noetherian $\mathbb{N}$-graded algebras which fit in an exact sequence as in Prop. 3.8.13. Then*

$$\operatorname{cd}\beta_* \leq \operatorname{cd}\alpha_* \leq \operatorname{cd}\beta_* + 1 \tag{3.41}$$

PROOF. It is easy to see that

$$\operatorname{cd}\alpha_* = \operatorname{cd}\underline{\alpha}_* = \operatorname{cd}\omega_{\mathcal{A}}$$

and similarly for β. Furthermore from (3.28)(3.29) it follows that

$$\operatorname{cd}\omega_{\mathcal{A}} = \max(\operatorname{cd}\tau_{\mathcal{A}} - 1, 0) \tag{3.42}$$

Combining this with Prop. 3.8.13(3) yields what we want. □

LEMMA 3.8.15. *Assume $\mathcal{A}$ is noetherian and $\mathbb{N}$-graded. Let $\mathcal{T}$ be a right bounded graded $\mathcal{A}-\mathcal{A}$-bimodule. Then for $\mathcal{M} \in \operatorname{Gr}(\mathcal{A})$ we have*

$$\underline{\mathcal{H}om}_{\mathcal{A}}(\mathcal{T}, \mathcal{M}) = \underline{\mathcal{H}om}_{\mathcal{A}}(\mathcal{T}, \tau(\mathcal{M})) \tag{3.43}$$

If $\mathcal{T}$ is in fact coherent as $o_X - o_X$-bimodule then $\underline{\mathcal{H}om}_{\mathcal{A}}(\mathcal{T}, \mathcal{M})$ is torsion.

PROOF. To prove (3.43) it is sufficient to show that the right and lefthand side of that equation represent the same functor. This is routine, using the fact that if $\mathcal{N} \in \operatorname{Gr}(\mathcal{A})$ then $\mathcal{N} \otimes_{\mathcal{A}} \mathcal{T}$ is torsion by lemma 3.7.1.2.

Now we prove the second part of the lemma. By the first part we may clearly assume that $\mathcal{M}$ is torsion. The coherentness of $\mathcal{T}$ also implies that $\mathcal{T}$ is left bounded.

We claim that $\mathcal{T}$ is coherent as a graded $\mathcal{A}$-bimodule. In the same way as in 3.1.6 it suffices to show that $- \otimes_{\mathcal{A}} \mathcal{T}$ preserves $\operatorname{gr}(\mathcal{A})$. Now since $\mathcal{A}$ is noetherian, any object in $\operatorname{gr}(\mathcal{A})$ has a presentation consisting of objects of the form $\mathcal{P} \otimes_{o_X} \mathcal{A}$ where the $\mathcal{P}$ are noetherian o_X-modules. We now use the fact that $- \otimes_{\mathcal{A}} \mathcal{T}$ is right exact.

Hence $\underline{\mathcal{H}om}_{\mathcal{A}}(\mathcal{T},-)$ commutes with direct limits. Therefore we may assume that $\mathcal{M}$ is right bounded. But $\underline{\mathcal{H}om}_{\mathcal{A}}(\mathcal{T},\mathcal{M})$ is contained in $\underline{\mathcal{H}om}_{o_X}(\mathcal{T},\mathcal{M})$ which is now clearly also right bounded. This proves what we want. □

We will need the following variant of Proposition 3.8.13

LEMMA 3.8.16. *Assume that we have an epimorphism* $f : \mathcal{A} \to \mathcal{B}$ *such that* $I = \ker f$ *is a coherent object in* $\operatorname{Bigr}(o_X - o_X)$. *Then for* $\mathcal{M} \in \operatorname{Gr}(\mathcal{B})$ *we have*

$$R^i\tau_{\mathcal{A}}(\mathcal{M}_{\mathcal{A}}) \cong (R^i\tau_{\mathcal{B}}(\mathcal{M}))_{\mathcal{A}} \tag{3.44}$$

Furthermore if $\mathcal{B}$ *satisfies* χ *then so does* $\mathcal{A}$ *and* $\operatorname{cd}\tau_{\mathcal{B}} = \operatorname{cd}\tau_{\mathcal{A}}$.

PROOF. It is clear that (3.44) is a special case of Proposition 3.8.10.

Assume that $\mathcal{B}$ satisfies χ. To show that $\mathcal{A}$ also satisfies χ we have to show that for $\mathcal{N} \in \operatorname{gr}(\mathcal{A})$ one has $R^i\tau_{\mathcal{A}}(\mathcal{N})_{\geq 0} \in \operatorname{gr}(\mathcal{A})$. Let $\mathcal{K} = \ker(\mathcal{N} \to \mathcal{N} \otimes_{\mathcal{A}} \mathcal{B})$. Since $\mathcal{N}$ is noetherian, the same is true for $\mathcal{K}$. Furthermore by Corollary 3.7.2 it follows that $\mathcal{K}$ is torsion. Hence

$$R\tau^i_{\mathcal{A}}(\mathcal{K}) = \begin{cases} \mathcal{K} & \text{if } i = 0 \\ 0 & \text{if } i > 0 \end{cases}$$

Thus the long exact sequence for $R^i\tau_{\mathcal{A}}$ together with (3.44) yields an exact sequence

$$0 \to \mathcal{K} \to \tau_{\mathcal{A}}(\mathcal{N}) \to \tau_{\mathcal{B}}(\mathcal{N} \otimes_{\mathcal{A}} \mathcal{B}) \to 0$$

and isomorphisms

$$R^i\tau_{\mathcal{A}}(\mathcal{N}) \to R^i\tau_{\mathcal{B}}(\mathcal{N} \otimes_{\mathcal{A}} \mathcal{B})$$

for $i \geq 1$. This yields that $\mathcal{A}$ satisfies χ and also that $\operatorname{cd}\tau_{\mathcal{B}} \geq \operatorname{cd}\tau_{\mathcal{A}}$. Since (3.44) implies that $\operatorname{cd}\tau_{\mathcal{B}} \leq \operatorname{cd}\tau_{\mathcal{A}}$, we are through. □

3.9. Higher inverse images

Assume that $\alpha : Y \to X$ is a morphism of quasi-schemes. As was observed above the definition of $R^i\alpha_*$ presents no difficulty since $\operatorname{Mod}(Y)$ has enough injectives. However the definition of $L_i\alpha^*$ is more delicate since we have not assumed that $\operatorname{Mod}(X)$ has enough flat objects. As an approximation we define $L_i\alpha^*$ as the functor $\operatorname{MOD}(X) \to \operatorname{MOD}(Y)$ given by

$$\mathcal{H}om_{o_Y}(L_i\alpha^*\mathcal{M}, E) = \mathcal{E}xt^i_{o_X}(\mathcal{M}, \alpha_* E)$$

where E runs through the injectives in $\operatorname{Mod}(Y)$. Furthermore we will define $\operatorname{cd}\alpha^*$ as the maximum i such that $L_i\alpha^*$, restricted to $\operatorname{Mod}(Y)$ is non-zero.

LEMMA 3.9.1. *Let* X *be a quasi-scheme and let* $\mathcal{A} \in \operatorname{Alg}(X)$. *Put* $Y = \operatorname{Spec}\mathcal{A}$ *and let* $\alpha : Y \to X$ *be the structure map. Then one has* $L_i\alpha^*(-) = \mathcal{T}or_i^{o_X}(-,\mathcal{A})$.

This lemma is proved in a similar way as the lemma below which covers the graded case. In the graded case we need more hypotheses since we have defined things in less generality.

LEMMA 3.9.2. *Let* X *be a noetherian quasi-scheme and let* $\mathcal{A}$ *be a noetherian* $\mathbb{N}$*-graded algebra on* X. *Put* $Y = \operatorname{Proj}\mathcal{A}$. *Assume that* $\mathcal{M} \in \operatorname{Mod}(X)$ *is such that* $\underline{\mathcal{T}or}_i^{o_X}(\mathcal{M},\mathcal{A}) \in \operatorname{Gr}(\mathcal{A})$. *Then*

$$L_i\alpha^*\mathcal{M} = \pi(\underline{\mathcal{T}or}_i^{o_X}(\mathcal{M},\mathcal{A}))$$

PROOF. We have to show that

$$\mathrm{Ext}^i_{o_X}(\mathcal{M}, \alpha_* F) = \mathrm{Hom}_{\mathrm{QGr}(\mathcal{A})}(\pi(\underline{\mathcal{T}or}_i^{o_X}(\mathcal{M}, \mathcal{A})), F) \tag{3.45}$$

where F runs through the injectives in $\mathrm{Mod}(Y)$. Put $F = \pi E$. It follows from lemma 3.7.3 that E is an injective in $\mathrm{Gr}(\mathcal{A})$ satisfying $\tilde{E} = E$. Then by adjointness the righthand side of (3.45) becomes

$$\mathrm{Ext}^i_{o_X}(\mathcal{M}, \mathcal{H}om_{\mathrm{Gr}(\mathcal{A})}(\mathcal{A}, E)) = \mathrm{Ext}^i_{o_X}(\mathcal{M}, E_0)$$

Hence the assertion we have to prove boils down to $\alpha_*\pi E = E_0$. This follows from the fact that $\alpha_*(-) = \omega(-)_0$. □

3.10. Algebras which are strongly graded modulo a Serre subcategory

The results in this section will be used in §3.11.

Let $\mathcal{D}$ be an abelian category and let $\mathcal{A} \in \mathrm{Gralg}(\mathcal{D})$. Put $\mathcal{Q}_n = \mathrm{coker}(\mathcal{A}_n \otimes_{\mathcal{D}} \mathcal{A}_{-n} \to \mathrm{id}_{\mathcal{D}})$. It is clear that $\mathcal{Q}_n \in \mathrm{Bimod}(\mathcal{D} - \mathcal{D})$. We call $\mathcal{A}$ *strongly graded* if $\mathcal{Q}_n = 0$ for all n. By copying the proof from the ring case [**26**, Thm. I.3.4, p16] one easily shows that the functors $(-)_0$ and $- \otimes_{\mathcal{A}_0} \mathcal{A}$ define inverse equivalences between $\mathrm{Gr}(\mathcal{A})$ and $\mathrm{Mod}(\mathcal{A}_0)$.

Let $\mathcal{S}$ be a Serre subcategory of $\mathrm{Mod}(\mathcal{A}_0)$. Define $\mathcal{S}(\mathcal{A})$ as the full subcategory of objects $\mathcal{T}$ in $\mathrm{Gr}(\mathcal{A})$ such that $\mathcal{T}_n$ is in $\mathcal{S}$ for every n. It is clear that $\mathcal{S}(\mathcal{A})$ is a Serre subcategory of $\mathrm{Gr}(\mathcal{A})$.

We say that $\mathcal{A}$ is strongly graded with respect to $\mathcal{S}$ if $- \otimes_{\mathcal{A}_0} \mathcal{Q}_n$ sends $\mathrm{Mod}(\mathcal{A}_0)$ to $\mathcal{S}$ and if $- \otimes_{\mathcal{A}_0} \mathcal{A}$ sends $\mathcal{S}$ to $\mathcal{S}(\mathcal{A})$.

Then one has the following:

LEMMA 3.10.1. *Assume that $\mathcal{A}$ is strongly graded with respect to $\mathcal{S}$. Then the functors $(-)_0$ and $- \otimes_{\mathcal{A}_0} \mathcal{A}$ factor over $\mathcal{S}(\mathcal{A})$ and $\mathcal{S}$ and in this way they define inverse equivalences between* $\mathrm{Gr}(\mathcal{A})/\mathcal{S}(\mathcal{A})$ *and* $\mathrm{Mod}(\mathcal{A}_0)/\mathcal{S}$.

PROOF. For the convenience of the reader we copy the proof of [**26**, Thm. I.3.4, p16] suitably modified.

If $\mathcal{M} \in \mathrm{Gr}(\mathcal{A})$ and if $\mathcal{P}_n$ is a submodule of $\mathcal{M}_n$ then we define $\mathcal{P}_n\mathcal{A}_m$ as the image of $\mathcal{P}_n \otimes_{\mathcal{D}} \mathcal{A}_m$ in $\mathcal{M}_{m+n}$.

We show that $\mathcal{M} = \mathcal{M}_0 \otimes_{\mathcal{A}_0} \mathcal{A}$ modulo $\mathcal{S}(\mathcal{A})$. Working modulo $\mathcal{S}$ we have the following.

$$\mathcal{M}_{m+n} = \mathcal{M}_{m+n}\mathcal{A}_0 = \mathcal{M}_{m+n}\mathcal{A}_{-m}\mathcal{A}_m \subset \mathcal{M}_n\mathcal{A}_m \subset \mathcal{M}_{m+n} \tag{3.46}$$

The second equality follows from

$$\begin{aligned}\mathcal{M}_{m+n}/\mathcal{M}_{m+n}\mathcal{A}_{-m}\mathcal{A}_m &= \mathrm{coker}(\mathcal{M}_{m+n} \otimes_{\mathcal{D}} \mathcal{A}_{-m} \otimes_{\mathcal{D}} \mathcal{A}_m \to \mathcal{M}_{m+n}) \\ &= \mathcal{M}_{m+n} \otimes_{\mathcal{D}} \mathcal{Q}_{-m}\end{aligned}$$

(3.46) yields $\mathcal{M}_n\mathcal{A}_m = \mathcal{M}_{m+n}$ modulo $\mathcal{S}$. Now we claim that the multiplication map

$$\mu : \mathcal{M}_0 \otimes_{\mathcal{A}_0} \mathcal{A} \to \mathcal{M}$$

is in fact an isomorphism modulo $\mathcal{S}(\mathcal{A})$. By what we have done so far μ is clearly epic modulo $\mathcal{S}(\mathcal{A})$. Let $\mathcal{K}$ be the kernel of μ. We have $\mathcal{K}_0 = 0$ and hence modulo $\mathcal{S}$, $\mathcal{K}_n = \mathcal{K}_0\mathcal{A}_n = 0$. Thus $\mathcal{K} \in \mathcal{S}(\mathcal{A})$ and we are done.

We can now finish the proof, provided we can show that the functor $- \otimes_{\mathcal{A}_0} \mathcal{A}$ is well defined on $\mathrm{Mod}(\mathcal{A}_0)/\mathcal{S}$. Inspection shows that we only have to check the following case : assume that $\mathcal{M} \to \mathcal{N}$ is injective in $\mathrm{Mod}(\mathcal{A}_0)$. Then $\mathcal{K} =$

$\ker(\mathcal{M} \otimes_{\mathcal{A}_0} \mathcal{A} \to \mathcal{N} \otimes_{\mathcal{A}_0} \mathcal{A})$ should be in $\mathcal{S}(\mathcal{A})$. Since $\mathcal{K}_0 = 0$ this follows from the previous discussion. □

3.11. The positive part of certain graded algebras.

In this section we fix the following situation. X is a noetherian quasi-scheme, $\mathcal{S}$ is a localizing subcategory in $\operatorname{Mod}(X)$. $\mathcal{A}$ is a noetherian algebra on X with $\mathcal{A}_0 = o_X$ which is strongly graded with respect to $\mathcal{S}$ (see §3.10). With $\mathcal{A}_{\geq 0}$ we denote the positive part of $\mathcal{A}$. It is easy to see that $\mathcal{A}_{\geq 0}$ is also noetherian.

We note the following.

LEMMA 3.11.1. *Let $\mathcal{M} \in \operatorname{Gr}(\mathcal{A}_{\geq 0})$. Then the $\mathcal{A}_{\geq 0}$-module structure on $\tilde{\mathcal{M}}$ extends in a natural way to an $\mathcal{A}$-module structure.*

PROOF. Multiplication defines graded $\mathcal{A}_{\geq 0}$-bimodule maps

$$\mathcal{A}_{\geq m} \otimes_{\mathcal{A}_{\geq 0}} \mathcal{A}_{\geq n} \to \mathcal{A}_{\geq m+n}$$

Applying $\underline{\mathcal{H}om}_{\mathcal{A}_{\geq 0}}(-, \mathcal{M})$ yields a map

$$\underline{\mathcal{H}om}_{\mathcal{A}_{\geq 0}}(\mathcal{A}_{\geq m+n}, \mathcal{M}) \to \underline{\mathcal{H}om}_{\mathcal{A}_{\geq 0}}(\mathcal{A}_{\geq m}, \underline{\mathcal{H}om}_{\mathcal{A}_{\geq 0}}(\mathcal{A}_{\geq n}, \mathcal{M}))$$

which by adjointness yields a map

$$\underline{\mathcal{H}om}_{\mathcal{A}_{\geq 0}}(\mathcal{A}_{\geq m+n}, \mathcal{M}) \otimes_{\mathcal{A}_{\geq 0}} \mathcal{A}_{\geq m} \to \underline{\mathcal{H}om}_{\mathcal{A}_{\geq 0}}(\mathcal{A}_{\geq n}, \mathcal{M})$$

Taking direct limits over n, using (3.23) and letting m go to $-\infty$ we find a map

$$\tilde{\mathcal{M}} \otimes_{\mathcal{A}_{\geq 0}} \mathcal{A} \to \tilde{\mathcal{M}}$$

A straightforward, but mildly tedious verification shows that this is an $\mathcal{A}$-module structure. □

According to Corollary 3.7.2 we have the following inverse equivalences

$$\operatorname{Gr}(\mathcal{A})/\operatorname{Tors}(\mathcal{A}) \underset{-\otimes_{\mathcal{A}_{\geq 0}}\mathcal{A}}{\overset{(-)_{\mathcal{A}_{\geq 0}}}{\rightleftarrows}} \operatorname{Gr}(\mathcal{A}_{\geq 0})/\operatorname{Tors}(\mathcal{A}_{\geq 0}) \tag{3.47}$$

By lemma 3.10.1 we have inverse equivalences

$$\operatorname{Mod}(\mathcal{A}_0)/\mathcal{S} \underset{(-)_0}{\overset{-\otimes_{\mathcal{A}_0}\mathcal{A}}{\rightleftarrows}} \operatorname{Gr}(\mathcal{A})/\mathcal{S}(\mathcal{A}) \tag{3.48}$$

In order to combine these equivalences we observe that $\operatorname{Tors}(\mathcal{A}) \subset \mathcal{S}(\mathcal{A})$. Indeed let $\mathcal{M} \in \operatorname{Tors}(\mathcal{A})$. Since $\operatorname{Tors}(\mathcal{A})$ and $\mathcal{S}(\mathcal{A})$ are closed under direct limits we may assume that $\mathcal{M}$ is right bounded. But then $\mathcal{M}(N)_0 = 0 \in \mathcal{S}$ for $N \gg 0$ and hence according to lemma 3.10.1 we have $\mathcal{M}(N) \in \mathcal{S}(\mathcal{A})$. Thus the same holds for $\mathcal{M}$.

Define $\operatorname{Q}\mathcal{S}(\mathcal{A}_{\geq 0})$ as the image in $\operatorname{QGr}(\mathcal{A}_{\geq 0})$ of $\mathcal{S}(\mathcal{A}_{\geq 0})$ under the quotient map. One notes that $\operatorname{Q}\mathcal{S}(\mathcal{A}_{\geq 0})$ has the following alternative description.

LEMMA 3.11.2. *$\operatorname{Q}\mathcal{S}(\mathcal{A}_{\geq 0})$ is precisely the image of $\mathcal{S}(\mathcal{A})$ under $\pi_{\mathcal{A}}(-)_{\mathcal{A}_{\geq 0}}$.*

PROOF. If we look at the commutative diagram

$$\begin{array}{ccc} \operatorname{Gr}(\mathcal{A}_{\geq 0}) & \xrightarrow{\pi_{\mathcal{A}_{\geq 0}}} & \operatorname{QGr}(\mathcal{A}_{\geq 0}) \\ {\scriptstyle (-)_{\mathcal{A}_{\geq 0}}}\uparrow & & {\scriptstyle (-)_{\mathcal{A}_{\geq 0}}}\uparrow \\ \operatorname{Gr}(\mathcal{A}) & \xrightarrow{\pi_{\mathcal{A}}} & \operatorname{QGr}(\mathcal{A}) \end{array}$$

then we see that it is sufficient to show that the image of $\mathcal{S}(\mathcal{A})$ under $(-)_{\mathcal{A}_{\geq 0}}$ is $\mathcal{S}(\mathcal{A}_{\geq 0})$ modulo $\mathrm{Tors}(\mathcal{A}_{\geq 0})$. It is clear that this image is indeed contained in $\mathcal{S}(\mathcal{A}_{\geq 0})$. Conversely let $\mathcal{M} \in \mathcal{S}(\mathcal{A}_{\geq 0})$. According to Corollary 3.7.2 we have $\mathcal{M} = (\mathcal{M} \otimes_{\mathcal{A}_{\geq 0}} \mathcal{A})_{\mathcal{A}_{\geq 0}}$ modulo $\mathrm{Tors}(\mathcal{A}_{\geq 0})$. Since $\mathcal{M} \otimes_{\mathcal{A}_{\geq 0}} \mathcal{A}$ is contained in $\mathcal{S}(\mathcal{A})$ we are through. □

Combining (3.47),(3.48) and the previous lemma yields equivalences.

$$\mathrm{Mod}(\mathcal{A}_0)/\mathcal{S} \underset{(-)_0}{\overset{-\otimes_{\mathcal{A}_0}\mathcal{A}}{\rightleftarrows}} \mathrm{Gr}(\mathcal{A})/\mathcal{S}(\mathcal{A}) \underset{-\otimes_{\mathcal{A}_{\geq 0}}\mathcal{A}}{\overset{(-)_{\mathcal{A}_{\geq 0}}}{\rightleftarrows}} \mathrm{QGr}(\mathcal{A}_{\geq 0})/\,\mathrm{Q}\mathcal{S}(\mathcal{A}_{\geq 0})$$

Looking only at the outer categories yields equivalences

$$\mathrm{Mod}(\mathcal{A}_0)/\mathcal{S} \underset{(-\otimes_{\mathcal{A}_{\geq 0}}\mathcal{A})_0}{\overset{-\otimes_{\mathcal{A}_0}\mathcal{A}_{\geq 0}}{\rightleftarrows}} \mathrm{QGr}(\mathcal{A}_{\geq 0})/\,\mathrm{Q}\mathcal{S}(\mathcal{A}_{\geq 0})$$

Define $U = \operatorname{Spec} \mathcal{A}_0$, $Y = \operatorname{Proj} \mathcal{A}_{\geq 0}$ and let the map $\alpha : Y \to U$ be given by $\alpha^* = \pi(- \otimes_{\mathcal{A}_0} \mathcal{A}_{\geq 0})$.

LEMMA 3.11.3. *Let $\mathcal{M} \in \mathrm{Gr}(\mathcal{A}_{\geq 0})$. Modulo $\mathcal{S}$, $\alpha_* \pi \mathcal{M}$ is given by $(\mathcal{M} \otimes_{\mathcal{A}_{\geq 0}} \mathcal{A})_0$.*

PROOF. Without loss of generality we may assume that $\mathcal{M} = \tilde{\mathcal{M}}$. Hence in particular by lemma 3.11.1 $\mathcal{M} = \mathcal{N}_{\mathcal{A}_{\geq 0}}$ where $\mathcal{N}$ is a graded $\mathcal{A}$-module. Furthermore $\alpha_* \pi \mathcal{M} = \mathcal{M}_0 = \mathcal{N}_0$.

According to Corollary 3.7.2 we now have that the canonical map

$$\mathcal{M} \otimes_{\mathcal{A}_{\geq 0}} \mathcal{A} = \mathcal{N}_{\mathcal{A}_{\geq 0}} \otimes_{\mathcal{A}_{\geq 0}} \mathcal{A} \to \mathcal{N}$$

is an isomorphism modulo $\mathrm{Tors}(\mathcal{A}) \subset \mathcal{S}(\mathcal{A})$. Hence the restricted map

$$(\mathcal{M} \otimes_{\mathcal{A}_{\geq 0}} \mathcal{A})_0 \to \mathcal{N}_0$$

is an isomorphism modulo $\mathcal{S}$. This is precisely what we had to prove. □

Let us now introduce the more suggestive notation $\alpha^{-1}(\mathcal{S})$ for $\mathrm{Q}\mathcal{S}(\mathcal{A}_{\geq 0})$. Summarizing everything, we have shown.

PROPOSITION 3.11.4. *Let $U, Y, \alpha, \mathcal{S}$ be as above. Then the functors α^*, α_* factor through $\mathcal{S}$ and $\alpha^{-1}(\mathcal{S})$ to define inverse equivalences between* $\mathrm{Mod}(U)/\mathcal{S}$ *and* $\mathrm{Mod}(Y)/\alpha^{-1}(\mathcal{S})$.

3.12. Veronese subalgebras

Let X be a quasi-scheme and let $\mathcal{A} \in \mathrm{Gralg}(X)$ be noetherian and $\mathbb{N}$-graded. The n'th Veronese of $\mathcal{A}$ is the graded subalgebra $\mathcal{A}^{(n)}$ of $\mathcal{A}$ defined by $(\mathcal{A}^{(n)})_m = \mathcal{A}_{nm}$. If $\mathcal{M} \in \mathrm{Gr}(\mathcal{A})$ then $\mathcal{M}^{(n)}$ is defined similarly.

Then we have the following.

LEMMA 3.12.1. *Assume that X and $\mathcal{A}$ are noetherian and that $\mathcal{A}$ is generated in degree one (cfr §3.8). Then the functors*

$$\mathrm{Gr}(\mathcal{A}) \to \mathrm{Gr}(\mathcal{A}^{(n)}) : \mathcal{M} \to \mathcal{M}^{(n)}$$

$$\mathrm{Gr}(\mathcal{A}^{(n)}) \to \mathrm{Gr}(\mathcal{A}) : \mathcal{N} \to \mathcal{N} \otimes_{\mathcal{A}^{(n)}} \mathcal{A}$$

factor over $\mathrm{Tors}(\mathcal{A})$ *and* $\mathrm{Tors}(\mathcal{A}^{(n)})$ *and in this way define inverse equivalences between* $\operatorname{Proj} \mathcal{A}$ *and* $\operatorname{Proj} \mathcal{A}^{(n)}$.

PROOF. This is formally similar to the ring case. See for example [**37**]. □

CHAPTER 4

Pseudo-compact rings

In the sequel we will study the formal local structure of some specific quasi-schemes. It turns out that this is best described in terms of pseudo-compact rings, so for the convenience of the reader we collect some of the properties of such rings here. Most of this is taken from [**36**] and [**16**]. We refer the reader to these papers for more information.

A topological right module M over a topological ring A is *pseudo-compact* if it is Hausdorf, complete and its topology is generated by right submodules of finite colength. A is said to be a *pseudo-compact ring* if A is pseudo-compact as a right A-module. Left pseudo-compact is defined similarly. By $\mathrm{PC}(A)$, we will denote the category of right pseudo-compact modules over a right pseudo-compact ring. By [**16**] $\mathrm{PC}(A)$ is an abelian category satisfying AB5* and AB3.

Let A be a pseudo-compact ring. The dual category of $\mathrm{PC}(A)$ is a locally finite category. That is a Grothendieck category generated by objects of finite length. Conversely, if $\mathcal{C}$ is a locally finite category then $\mathcal{C}$ can be realized as $\mathrm{PC}(A)^\circ$ for some pseudo-compact ring A [**16**]. A is constructed as follows. Let E be an injective cogenerator of $\mathcal{C}$, containing every indecomposable injective at least once and put $A = \mathrm{End}_{\mathcal{C}}(E)^\circ$. If S is a finite length subobject of E then

$$\mathfrak{l}(S) = \{f \in A \mid f(S) = 0\}$$

defines a right ideal of finite colength in A. We take these right ideal as the basis for a topology on A. In this way A becomes a pseudo-compact ring. If $M \in \mathcal{C}$ then we topologize $\mathrm{Hom}_{\mathcal{C}}(M, E)$ in a similar way. The functor $M \mapsto \mathrm{Hom}_{\mathcal{C}}(M, E)$ defines a duality between $\mathcal{C}$ and $\mathrm{PC}(A)$.

As an application we obtain :

LEMMA 4.1. *Inverse limits of projectives are projective in* $\mathrm{PC}(A)$.

PROOF. This follows from the fact that $\mathrm{PC}(A)^\circ$ is a locally finite category and hence direct limits of injectives are injective. □

In nice cases there are good relations between the properties of $\mathrm{PC}(A)$ and $\mathrm{Mod}(A)$. For example by [**16**] it follows that the forgetful functor $\mathrm{PC}(A) \to \mathrm{Mod}(A)$ is faithful and commutes with kernels, cokernels and products. By [**36**, Lemma 3.4], an object in $\mathrm{PC}(A)$ is simple in $\mathrm{PC}(A)$ if and only it is simple in $\mathrm{Mod}(A)$. A similar result holds for the property of being noetherian [**36**, Cor 3.10]. As usual we denote by $\mathrm{pc}(A)$ the category of noetherian pseudo-compact A-modules.

An object M in $\mathrm{PC}(A)$ is said to be finitely generated in $\mathrm{PC}(A)$ if M is a quotient of A^n in $\mathrm{PC}(A)$ for some n. If $M, N \in \mathrm{PC}(A)$, M finitely generated then according to [**36**, Prop. 3.5] we have

$$\mathrm{Hom}_{\mathrm{PC}(A)}(M, N) = \mathrm{Hom}_{\mathrm{Mod}(A)}(M, N)$$

In particular, M is finitely generated in $\mathrm{PC}(A)$ if and only if it is finitely generated in $\mathrm{Mod}(A)$. From this one can deduce :

LEMMA 4.2. [**36**, Cor. 3.8] *If M is a finitely generated pseudo-compact A-module then $L \subset M$ is open if and only if M/L is of finite length and pseudo-compact when equipped with the discrete topology.*

(Note that a linear Hausdorf topology on a module of finite length is automatically discrete.) If A is noetherian then it follows from [**36**, Prop. 3.19] that the forgetful functor $\mathrm{pc}(A) \to \mathrm{mod}(A)$ is an equivalence of categories. More generally we say that A is *locally noetherian* if for every primitive idempotent e in A we have that eA is noetherian. Assume that A is locally noetherian. Then by [**36**, Cor. 3.15] we have that the forgetful functor $\mathrm{pc}(A) \to \mathrm{mod}(A)$ is fully faithful and closed under extensions. Let $\mathrm{PCFin}(A)$ denote the category of finite length objects in $\mathrm{PC}(A)$. It follows that the objects are precisely the finite length objects in $\mathrm{Mod}(A)$ whose Jordan-Holder quotients are pseudo-compact simples. It also follows that if M is a noetherian A-module then the topology on M is simply the cofinite topology.

Let A be an arbitrary pseudo-compact ring. Then we denote by $\mathrm{Dis}(A)$ the category of topological A-modules which are discrete. It is clear that $\mathrm{Dis}(A)$ is the full subcategory of $\mathrm{Mod}(A)$ consisting of modules M such that for all $m \in M$, $\mathrm{Ann}_A(m)$ is open in A, or equivalently that mA is pseudo-compact of finite length (this follows for example from lemma 4.2). From this we deduce that $\mathrm{Dis}(A)$ is a locally finite category. Clearly $\mathrm{PC}(A) \cap \mathrm{Dis}(A) = \mathrm{PCFin}(A)$ (where the intersection is taken inside the category of topological right A-modules).

It is interesting to observe that since $\mathrm{Dis}(A)$ is locally finite there must necessarily exist another pseudo-compact ring A^* such that $\mathrm{PC}(A^*) = \mathrm{Dis}(A)^\circ$. In nice cases we have $A^* = A^\circ$ (see for example Proposition 4.10 below).

Let $\mathrm{Top}(A)$ be the additive category of topological right A-modules. For $M, N \in \mathrm{Top}(A)$ we have functors

$$\mathrm{Hom}_{\mathrm{Top}(A)}(M, -) : \mathrm{Dis}(A) \to \mathbf{Ab} \tag{4.1}$$

$$\mathrm{Hom}_{\mathrm{Top}(A)}(-, N) : \mathrm{PC}(A)^\circ \to \mathbf{Ab} \tag{4.2}$$

PROPOSITION 4.3. *Let $(M_i)_{i\in J}$, $(N_j)_{j\in J}$ be respectively an inverse system in $\mathrm{PC}(A)$ and a directed system in $\mathrm{Dis}(A)$. Then*

$$\mathrm{Hom}_{\mathrm{Top}(A)}(\varprojlim_i M_i, \varinjlim_j N_j) = \varinjlim_i \varinjlim_j \mathrm{Hom}_{\mathrm{Top}(A)}(M_i, N_j) \tag{4.3}$$

PROOF. If $M \in \mathrm{PC}(A)$, $N \in \mathrm{Dis}(A)$ then every continuous morphism $f : M \to N$ has an open kernel M'. This means that $N' = \operatorname{im} f \cong M/M'$ has finite length. Thus we have the following equalities

$$\mathrm{Hom}_{\mathrm{Top}(A)}(M, N) = \varinjlim_{N'} \mathrm{Hom}_{\mathrm{PC}(A)}(M, N') \tag{4.4}$$

$$= \varinjlim_{M'} \mathrm{Hom}_{\mathrm{Dis}(A)}(M/M', N) \tag{4.5}$$

$$= \varinjlim_{M',N'} \mathrm{Hom}_A(M/M', N') \tag{4.6}$$

where M' runs trough the open submodules in M and N' runs through the finite length submodules of N.

Now let us for example show that $\operatorname{Hom}_{\operatorname{Top}(A)}(-,N)$ sends inverse limits to direct limits. We have

$$\begin{aligned}
\operatorname{Hom}_{\operatorname{Top}(A)}(\varprojlim_i M_i, N) &= \varinjlim_{N'} \operatorname{Hom}_{\operatorname{PC}(A)}(\varprojlim_i M_i, N') \qquad \text{(eq. (4.4))}\\
&= \varinjlim_{N'} \operatorname{Hom}_{\operatorname{PC}(A)^\circ}(N', \varinjlim_i M_i)\\
&= \varinjlim_{i,N'} \operatorname{Hom}_{\operatorname{PC}(A)^\circ}(N', M_i)\\
&= \varinjlim_{i,N'} \operatorname{Hom}_{\operatorname{PC}(A)}(M_i, N')\\
&= \varinjlim_{i} \operatorname{Hom}_{\operatorname{Top}(A)}(M_i, N) \qquad \text{(eq. (4.4))}
\end{aligned}$$

The third equality follows from the fact that N' is a noetherian object in the locally noetherian category $\operatorname{PC}(A)^\circ$.

The proof that $\operatorname{Hom}_{\operatorname{Top}(A)}(M,-)$ commutes with direct limits is similar. □

Since $\operatorname{Dis}(A)$ has enough injectives and $\operatorname{PC}(A)$ has enough projectives, we can define the derived functors of (4.1) and (4.2). Let us temporarily denote them by $\operatorname{Ext}^i_I(M,-)$ and $\operatorname{Ext}^i_{II}(-,N)$.

LEMMA 4.4. *Ext^i_I and $\operatorname{Ext}^i_{II}$ coincide when both are defined. That is if $M \in \operatorname{PC}(A)$ and $N \in \operatorname{Dis}(A)$ then*

$$\operatorname{Ext}^i_I(M,N) = \operatorname{Ext}^i_{II}(M,N)$$

PROOF. To prove this we have to show that if P is projective in $\operatorname{PC}(A)$ then $\operatorname{Hom}_{\operatorname{Top}(A)}(P,-)$ is exact on $\operatorname{Dis}(A)$ and if E is an injective in $\operatorname{Dis}(A)$ then $\operatorname{Hom}_{\operatorname{Top}(A)}(-,E)$ is exact when evaluated on $\operatorname{PC}(A)$. Since these are obviously dual statements we only prove the first one.

By Proposition 4.3

$$\operatorname{Hom}_{\operatorname{Top}(A)}(P,N) = \varinjlim_{N'} \operatorname{Hom}_{\operatorname{PC}(A)}(P,N')$$

where N' runs through the finite length submodules of N. Since $\operatorname{Hom}_{\operatorname{PC}(A)}(P,-)$ is exact on $\operatorname{PC}(A)$ and $\varinjlim$ is exact on **Ab** we have to show that every exact sequence in $\operatorname{Dis}(A)$

$$0 \to N_1 \to N \to N_2 \to 0$$

can be obtained as a direct limit of exact sequences of the form

$$0 \to N_1' \to N' \to N_2' \to 0$$

where N_1', N_2', N' are finite length subobjects of N_1, N_2, N. That this is true follows easily from the fact that $\operatorname{Dis}(A)$ is locally finite. □

Henceforth we neglect the distinction between Ext^i_I and $\operatorname{Ext}^i_{II}$, and we simply write $\operatorname{Ext}^i_{\operatorname{Top}(A)}$. We then obtain the following generalization of Proposition 4.3

PROPOSITION 4.5. *Let $(M_i)_{i\in I}$, $(N_j)_{j\in J}$ be respectively an inverse system in $\operatorname{PC}(A)$ and a directed system in $\operatorname{Dis}(A)$. Then*

$$\operatorname{Ext}^i_{\operatorname{Top}(A)}(\varprojlim_i M_i, \varinjlim_j N_j) = \varinjlim_i \varinjlim_j \operatorname{Ext}^i_{\operatorname{Top}(A)}(M_i, N_j) \tag{4.7}$$

PROOF. Let $N \in \mathrm{Dis}(A)$ and let $E^{\cdot}$ be an injective resolution of N in $\mathrm{Dis}(A)$. Then

$$\begin{aligned}
\mathrm{Ext}^i_{\mathrm{Top}(A)}(\varprojlim_i M_i, N) &= H^i(\mathrm{Hom}_{\mathrm{Top}(A)}(\varprojlim_i M_i, E^{\cdot})) \\
&= \varinjlim_i H^i(\mathrm{Hom}_{\mathrm{Top}(A)}(M_i, E^{\cdot})) \qquad \text{(Proposition 4.3)} \\
&= \varinjlim_i \mathrm{Ext}^i(M_i, N)
\end{aligned}$$

The fact that $\mathrm{Ext}^i_{\mathrm{Top}(A)}(M, -)$ is compatible with direct limits is proved similarly. □

Of course the ordinary "Ext" in the abelian categories $\mathrm{PC}(A)$ and $\mathrm{Dis}(A)$ is also defined. Since $\mathrm{PC}(A)$ has enough projectives we clearly have

$$\mathrm{Ext}^i_{\mathrm{PC}(A)}(M, N) = \mathrm{Ext}^i_{\mathrm{Top}(A)}(M, N)$$

if $M, N \in \mathrm{PC}(A)$. Similarly if $M, N \in \mathrm{Dis}(A)$ then

$$\mathrm{Ext}^i_{\mathrm{Dis}(A)}(M, N) = \mathrm{Ext}^i_{\mathrm{Top}(A)}(M, N)$$

since $\mathrm{Dis}(A)$ has enough injectives.

PROPOSITION 4.6. *Let $M \in \mathrm{PC}(A)$, $N \in \mathrm{Dis}(A)$. We have the following formulas*

$$\operatorname{proj\,dim}_{\mathrm{PC}(A)} M = \sup_i \{i \mid \mathrm{Ext}^i_{\mathrm{PC}(A)}(M, -) \neq 0\}$$

$$\operatorname{inj\,dim}_{\mathrm{Dis}(A)} N = \sup_i \{i \mid \mathrm{Ext}^i_{\mathrm{Dis}(A)}(-, N) \neq 0\}$$

If A is locally noetherian and if M is noetherian in $\mathrm{PC}(A)$ *then*

$$\operatorname{proj\,dim}_{\mathrm{PC}(A)} M = \sup_i \{i \mid \exists S \text{ simple } : \mathrm{Ext}^i_{\mathrm{PC}(A)}(M, S) \neq 0\}$$

and if A^* is locally noetherian and N is artinian in $\mathrm{Dis}(A)$ *then*

$$\operatorname{inj\,dim}_{\mathrm{Dis}(A)} N = \sup_i \{i \mid \exists T \text{ simple } : \mathrm{Ext}^i_{\mathrm{Dis}(A)}(T, N) \neq 0\}$$

PROOF. This is entirely classical. Let us for example prove the fourth equality. By degree shifting this amounts to showing that if $\mathrm{Ext}^1(T, N) = 0$ for all simple T and if N is artinian then N is injective.

Let E be the injective hull of N and $U = E/N$. Let $q : E \to U$ be the quotient map. By hypotheses E and hence U is artinian. Also by hypotheses, the restriction of q to $\mathrm{Soc}(E) \to \mathrm{Soc}(U)$ is a epimorphism. Since socles are by definition semisimple this last map has a splitting which can be lifted to a map $t : U \to E$. By hypotheses $s = qt$ is the identity on $\mathrm{Soc}(U)$, from which it follows that s is injective. Also $s^n(U)$ is a descending chain of submodules in U and thus $s^n(U) = s^{n+1}(U)$. The injectivity of s yields that $U = sU$ and thus s is an automorphism of U. Now ts^{-1} is a splitting of q and hence N is a direct summand of E, whence injective. □

As usual we define

$$\operatorname{gl\,dim} \mathrm{PC}(A) = \sup_{M \in \mathrm{PC}(A)} \operatorname{proj\,dim} M$$

$$\operatorname{gl\,dim} \mathrm{Dis}(A) = \sup_{N \in \mathrm{Dis}(A)} \operatorname{inj\,dim} N$$

COROLLARY 4.7. *The following holds.*

1. $\operatorname{gl\,dim} \operatorname{PC}(A)$ *is equal to the supremum of the projective dimensions of the simple pseudo-compact A-modules (as objects in* $\operatorname{PC}(A)$*).*
2. $\operatorname{gl\,dim} \operatorname{Dis}(A)$ *is equal to the supremum of the injective dimensions of the simple pseudo-compact A-modules (as objects in* $\operatorname{Dis}(A)$*).*
3. *If A and A^* are locally noetherian then* $\operatorname{gl\,dim} \operatorname{PC}(A)$ *and* $\operatorname{gl\,dim} \operatorname{Dis}(A)$ *are both equal to the supremum of the $i \in \mathbb{N}$ such that there exist simple pseudo-compact-A-modules S, T such that* $\operatorname{Ext}^i(S,T) \neq 0$.

PROOF. 1. and 2. are true by [**36**, Lemma 5.1]. 3. follows from 1. and 2. and the foregoing proposition. □

An object in $\operatorname{PC}(A)$ is said to be cosemisimple if it is a direct product of simple modules. This is equivalent to being semisimple in the dual category $\operatorname{PC}(A)^\circ$. If $M \in \operatorname{PC}(A)$ then we define $M/\operatorname{rad}(M)$ as the quotient of M which is the socle of M in $\operatorname{PC}(A)^\circ$. By construction $M/\operatorname{rad}(M)$ is the largest cosemisimple object in $\operatorname{PC}(A)$ which is a quotient of M. The *radical* of M, denoted by $\operatorname{rad}(M)$, is defined as the kernel of $M \to M/\operatorname{rad}(M)$. Since it is closed, it is pseudo-compact. From the fact that taking socles is left exact it also follows that the functor $M \mapsto M/\operatorname{rad}(M)$ is right exact. It is shown in [**16**] that $\operatorname{rad}(A)$ is a twosided ideal and coincides with the ordinary Jacobson radical of A. From the fact that $\operatorname{rad}(A)$ annihilates all cosemisimple objects in $\operatorname{PC}(A)$ we obtain

$$M \operatorname{rad}(A) \subset \operatorname{rad}(M) \tag{4.8}$$

LEMMA 4.8. *(Nakayama's lemma) If $M \in \operatorname{PC}(A)$ then $M = \operatorname{rad}(M)$ if and only if $M = 0$.*

PROOF. It is easy to see that a non-zero object in $\operatorname{PC}(A)$ maps onto at least one simple object. This proves the lemma. □

LEMMA 4.9. *The following are equivalent.*

1. *M is finitely generated.*
2. *$M/M\operatorname{rad}(A)$ is a finitely generated $A/\operatorname{rad}(A)$-module.*
3. *$M/\operatorname{rad}(M)$ is a finitely generated $A/\operatorname{rad}(A)$-module.*

If any one of these conditions holds then $\operatorname{rad}(M) = M\operatorname{rad}(A)$.

PROOF. It is clear that 1. implies 2. From (4.8) it follows that 2. implies 3. Hence we have to show that 3. implies 1. By lifting the generators of $M/\operatorname{rad}(M)$ we can construct a map $\theta : A^k \to M$ which becomes surjective after applying the functor $T \mapsto T/\operatorname{rad}(T)$. Given the right exactness of this functor we obtain that $C = \operatorname{rad}(C)$ for $C = \operatorname{coker}\theta$. By Nakayama it follows that $C = 0$. This proves the first part of the lemma.

If condition 2. holds then $M/M\operatorname{rad}(A)$ is a quotient of $(A/\operatorname{rad}(A))^k$ for some k, and hence $M/M\operatorname{rad}(A)$ is cosemisimple. We conclude that $\operatorname{rad}(M) \subset M\operatorname{rad}(A)$. □

For further reference we state the following formula

$$M/\operatorname{rad}(M) = \prod_{S \text{ simple}} S^{\alpha_{M,S}} \tag{4.9}$$

where $\alpha_{M,S} = \dim_{\operatorname{End}_{\operatorname{PC}(A)}(S)} \operatorname{Hom}_{\operatorname{PC}(A)}(M,S)$. This is easily proved by looking at the dual statement in $\operatorname{PC}(A)^\circ$.

The definition of a pseudo-compact ring is essentially onesided, which is somewhat inconvenient. We now introduce a symmetric notion.

Let k be a field. A pseudo-compact ring which is a k algebra is said to be *cofinite* if all simple pseudo-compact A-modules are finite dimensional over k.

PROPOSITION 4.10. *Assume that A is cofinite. Then*

1. $A/\operatorname{rad}(A) = \prod_{i\in I} M_{n_i}(D_i)$, *for finite dimensional division algebras* $(D_i)_{i\in I}$.
2. *The topology on A is generated by twosided ideals of finite codimension.*
3. *A is left and right pseudo-compact.*
4. $S \mapsto \operatorname{Hom}_k(S,k)$ *defines a duality between left and right pseudo-compact A-modules of finite length. In particular we can take* $A^* = A^\circ$.
5. *(Matlis-duality)* $M \mapsto \operatorname{Hom}_{\operatorname{Top}(k)}(M,k)$ *defines a duality between pseudo-compact left (right) A-modules and discrete right (left) A-modules. Thus* $\operatorname{PC}(A)^\circ = \operatorname{Dis}(A^\circ)$ *and* $\operatorname{Dis}(A)^\circ = \operatorname{PC}(A^\circ)$.

PROOF.
1. According to Gabriel $A/\operatorname{rad}(A)$ is a product of endomorphism rings of vectorspaces over division algebras. The simple pseudo-compact modules over such a ring will be finite dimensional if and only if $A/\operatorname{rad}(A)$ has the indicated form.
2. Let $L \subset A$ be an open ideal. Then $S = A/L$ is a pseudo-compact right A-module of finite length. Put $T = \operatorname{End}_k(S)$ and consider T as an A-bimodule in the obvious way. Then as right A-module, we have $T = S^{\dim_k S}$ and in particular T is pseudo-compact. There is a canonical map of A-bimodules $A \to T$, given by the right action of A on S. Let M be the kernel of this map. Then M is an open twosided ideal contained in L.
3. Since the topology on A is generated by twosided ideals of finite codimension (by 2.) we see that A is also pseudo-compact on the left.
4. This follows again easily from the fact that S is annihilated by an open twosided ideal.
5. This is a consequence of 3. and the fact that objects in $\operatorname{Dis}(A)$ are direct limits of finite length pseudo-compact modules and objects in $\operatorname{PC}(A)$ are inverse limits of finite length pseudo-compact modules. □

COROLLARY 4.11. *For a cofinite pseudo-compact k-algebra which is left and right locally noetherian the numbers* $\operatorname{gl}\dim \operatorname{PC}(A)$, $\operatorname{gl}\dim \operatorname{Dis}(A)$, $\operatorname{gl}\dim \operatorname{PC}(A^\circ)$ *and* $\operatorname{gl}\dim \operatorname{Dis}(A^\circ)$ *all coincide. We call this common value the global dimension of A and denote it by* $\operatorname{gl}\dim A$.

PROOF. This follows from Corollary 4.7.3 and Proposition 4.10.5. □

DEFINITION 4.12. Let A, B be cofinite k-algebras. Let M be a topological A-B-bimodule. Then M is bi-pseudo-compact if the topology on M is Hausdorff and complete and is generated by subbimodules $M' \subset M$ of finite codimension.

It is easy to see that bi-pseudo-compact bimodules form an abelian category satisfying AB3 and AB5*. We denote this category by $\operatorname{PC}(A-B)$.

If A, B, C are three cofinite algebras and $M \in \operatorname{PC}(A-B)$, $N \in \operatorname{PC}(B-C)$ then we define

$$M \mathbin{\hat{\otimes}_B} N = \varprojlim_{M',N'} M/M' \otimes_B N/N'$$

where M', N' run through all open subbimodules in M, N. By construction $M \hat{\otimes}_B N \in \mathrm{PC}(A - C)$. There are analogous definitions when M or N are onesided pseudo-compact modules.

One easily obtains

LEMMA 4.13. *$- \hat{\otimes}_B -$ is right exact and commutes with inverse limits in both its factors.*

The following is also standard.

LEMMA 4.14. *If M is finitely presented as right B-module then*

$$M \hat{\otimes}_B N = M \otimes_B N$$

If A, B are cofinite k-algebras then $A \hat{\otimes}_k B$ carries a canonical ring structure which makes it into a cofinite k-algebra. One has $\mathrm{PC}(A - B) = \mathrm{PC}(A^\circ \hat{\otimes}_k B)$. Furthermore the pseudo-compact simple $A^\circ \hat{\otimes} B$-modules are of the form $S \otimes_k T$ with S simple in $\mathrm{PC}(A)$ and T simple in $\mathrm{PC}(B)$ (this follows from the corresponding statement for finite dimensional algebras). From this one deduces $A \hat{\otimes}_k B / \operatorname{rad}(A \hat{\otimes}_k B) = A/\operatorname{rad}(A) \hat{\otimes}_k B/\operatorname{rad}(B)$.

This observation allows one to prove

LEMMA 4.15. *The forgetful functor $\mathrm{PC}(A - B) \to \mathrm{PC}(B)$ preserves projectives.*

PROOF. It suffices to prove this for the functor $\mathrm{PC}(A^\circ \hat{\otimes} B) \to \mathrm{PC}(B)$. By the above discussion all projectives in $\mathrm{PC}(A^\circ \hat{\otimes} B)$ are products of $Ae \hat{\otimes} fB$ where e, f are primitive idempotents in A, B. These are direct summands of $A \hat{\otimes} B$, whence it suffices to show that $(A \hat{\otimes} B)_B$ is projective. This follows from the fact that

$$\begin{aligned} A \hat{\otimes} B &= \varprojlim_{I,J} (A/I) \otimes (B/J) \\ &= \varprojlim_{I} (\varprojlim_{J} (A/I) \otimes (B/J)) \\ &= \varprojlim_{I} A/I \otimes B \end{aligned}$$

where I, J run through the open twosided ideals in A and B. Hence $(A \hat{\otimes} B)_B$ is an inverse limit of projective B-modules, and is therefore itself projective (lemma 4.1) □

We now give a structure theorem on cofinite k-algebras, which was more or less proved in [**36**] (in slightly greater generality). Below let $(e_i)_{i \in I}$ be a summable set of primitive idempotents in a pseudo-compact ring A, having sum 1 (as in [**16**]).

PROPOSITION 4.16. *Let A be cofinite, let M be a pseudo-compact A-module and let N be a bipseudo-compact A-bimodule. Put $A_{ij} = e_i A e_j$, $M_i = M e_i$, $N_{ij} = e_i N e_j$, equipped with the induced topology. Then*

$$A = \prod_{i,j} A_{ij} \tag{4.10}$$

$$M = \prod_{i} M_i \tag{4.11}$$

$$N = \prod_{i,j} e_i N e_j \tag{4.12}$$

These equalities are in fact homeomorphisms if we equip the righthand sides with the product topology. Furthermore the A_i are cofinite with $A_i/\operatorname{rad}(A_i) = D_i$ (where D_i is as in Proposition 4.10), the M_i are pseudo-compact A_i-modules and the A_{ij} and N_{ij} are bipseudo-compact $A_i - A_j$-modules.

PROOF. Similar to [**36**, Prop. 4.3]. □

In the case of a cofinite k-algebra, there is a simple test for a pseudo-compact module to be finitely generated.

LEMMA 4.17. *Assume that A is cofinite. Then $M \in \mathrm{PC}(A)$ is finitely generated if and only if for every simple pseudo-compact S we have that the dimension of $\mathrm{Hom}_{\mathrm{PC}(A)}(M,S)$ as a vector space over $\mathrm{End}_{\mathrm{PC}(A)}(S)$ is finite and is bounded independently of S.*

PROOF. This follows from Lemma 4.9 and (4.9). □

The following is a generalization of lemma 4.17.

LEMMA 4.18. *Assume that A is cofinite. Let $M \in \mathrm{PC}(A)$. Then M has a free resolution of length n in $\mathrm{PC}(A)$ (or, equivalently, in $\mathrm{Mod}(A)$)*

$$F_n \to F_{n-1} \to \cdots \to F_1 \to F_0 \to M \to 0$$

where F_i is of finite rank over A, if and only if for every $i \in \{0, \dots, n\}$ and for every simple S the dimension of $\mathrm{Ext}^i_{\mathrm{PC}(A)}(M,S)$ is finite and bounded independently of S.

PROOF. This is easily proved from the case $n = 0$, by degree shifting. □

We will need the following result.

PROPOSITION 4.19. *Assume that A is cofinite, $M \in \mathrm{PC}(A - A)$ and $S \in \mathrm{Dis}(A)$. Then $\mathrm{Ext}^i_{\mathrm{Top}(A)}(M,S) \in \mathrm{Dis}(A)$. Here the "Ext" is taken with respect to the right A-structure on M.*

PROOF. Let $P^\cdot$ be a projective resolution of M in $\mathrm{PC}(A - A)$. According to lemma 4.15 the terms of $P^\cdot$ are projective in $\mathrm{PC}(A)$. Hence we find

$$\mathrm{Ext}^i_{\mathrm{PC}(A)}(M,S) = H^i(\mathrm{Hom}_{\mathrm{PC}(A)}(P^\cdot, S))$$

Hence it suffices to show that if P is a projective object in $\mathrm{PC}(A - A)$ then $\mathrm{Hom}_{\mathrm{PC}(A)}(P,S) \in \mathrm{Dis}(A)$. Such a projective object is a product of direct summands of $A^\circ \hat{\otimes} A$, and since $\mathrm{Hom}_{\mathrm{PC}(A)}(-,S)$ transforms products in the first argument into direct sums (according to Proposition 4.3) it suffices to show that $\mathrm{Hom}_{\mathrm{PC}(A)}(A \hat{\otimes} A, S)$ lies in $\mathrm{Dis}(A)$. Now we have

$$\begin{aligned}
\mathrm{Hom}_{\mathrm{PC}(A)}(A \hat{\otimes} A, S) &= \mathrm{Hom}_{\mathrm{PC}(A)}(\varprojlim_I A/I \otimes A, S) \\
&= \varinjlim_I \mathrm{Hom}_{\mathrm{PC}(A)}(A/I \otimes A, S) \qquad \text{(Proposition 4.3)} \\
&= \varinjlim_I (A/I)^* \otimes S
\end{aligned}$$

Here I runs through the open twosided ideals in A. The right A-structure on $(A/I)^* \otimes S$ we use is the one on $(A/I)^*$. Thus $(A/I)^* \otimes S = (A/I)^t$ for some t, and we are done. □

We can define the derived functors of $-\hat{\otimes}_B-$ in both arguments. For lack of a better notation we denote them by $\mathrm{Tor}_i^{\mathrm{PC}(B)}(-,-)$. Thus if A, B, C are cofinite k-algebras and $M \in \mathrm{PC}(A-B)$, $N \in \mathrm{PC}(B-C)$ then we compute $\mathrm{Tor}_i^{\mathrm{PC}(B)}(M,N)$ in the usual way. For example we can start with a projective resolutions of M in $\mathrm{PC}(A-B)$. According to lemma 4.15 this yields a projective resolution of M in $\mathrm{PC}(B)$. We can also start with a projective resolution of N and get the same result.

We will need the following.

LEMMA 4.20. *Let E be an injective object in* $\mathrm{Dis}(C)$. *Then*

$$\mathrm{Hom}_{\mathrm{Top}(C)}(\mathrm{Tor}_i^{\mathrm{PC}(B)}(M,N),E) = \mathrm{Ext}^i_{\mathrm{Top}(B)}(M, \mathrm{Hom}_{\mathrm{Top}(C)}(N,E))$$

PROOF. This follows easily if we replace M by a projective resolution. □

CHAPTER 5

Cohen-Macaulay curves embedded in quasi-schemes

5.1. Preliminaries

In the rest of this paper k will be an algebraically closed base field. We will (usually tacitly) assume that all quasi-schemes are in $\operatorname{Qsch}/\operatorname{Spec} k$. Note that if $(X,\gamma) \in \operatorname{Qsch}/\operatorname{Spec} k$ then X contains the canonical object $\mathcal{O}_X = \gamma^* k$ (the "structure sheaf"). However this extra structure on X will not be used until §6.6.

Below $i : Y \to X$ will be a closed embedding of a commutative Cohen-Macaulay curve Y/k as a divisor (in the enriched sense) in a noetherian quasi-scheme X/k (§3.6). In the commutative case this hypothesis would imply that X is a surface in a neighborhood of Y.

Throughout this paper we will impose the following smoothness condition on X.

HYPOTHESIS (*). *Every object in* $\operatorname{Mod}(Y)$ *has finite injective dimension in* $\operatorname{Mod}(X)$.

It is easy to see that this is equivalent to the seemingly weaker condition.

HYPOTHESIS (*'). *For every* p *one has that* $\mathcal{O}_p$ *has finite injective dimension in* $\operatorname{Mod}(X)$.

The latter condition is sometimes automatic as can be seen from the following lemma.

LEMMA 5.1.1. *Assume that* Y *is smooth at* p. *Then* $\mathcal{O}_p$ *has finite injective dimension in* $\operatorname{Mod}(X)$.

PROOF. It is easy to see that we have to show that there is some n such that $\operatorname{Ext}^n_{\operatorname{Mod}(X)}(\mathcal{F}, \mathcal{O}_p) = 0$ for all $\mathcal{F} \in \operatorname{mod}(X)$. Using the long exact sequence for Ext we only have to show this in the following two cases.

1. The canonical map $\mathcal{F}(-Y) \to \mathcal{F}$ is monic.
2. $\mathcal{F} \in \operatorname{mod}(Y)$.

The lemma now follows from Proposition 5.1.2 below. □

PROPOSITION 5.1.2. *Let* $\mathcal{F} \in \operatorname{Mod}(X)$, $\mathcal{S} \in \operatorname{Mod}(Y)$. *Then*

1. *If the canonical map* $\mathcal{F}(-Y) \to \mathcal{F}$ *is monic then*
$$\operatorname{Ext}^i_{\operatorname{Mod}(X)}(\mathcal{F}, \mathcal{S}) = \operatorname{Ext}^i_{\operatorname{Mod}(Y)}(\mathcal{F}/\mathcal{F}(-Y), \mathcal{S})$$
2. *If* $\mathcal{F} \in \operatorname{Mod}(Y)$ *then there is a long exact sequence*

$$\to \operatorname{Ext}^{i-2}_{\operatorname{Mod}(Y)}(\mathcal{F}, \mathcal{S}(Y)) \to \operatorname{Ext}^i_{\operatorname{Mod}(Y)}(\mathcal{F}, \mathcal{S}) \to \operatorname{Ext}^i_{\operatorname{Mod}(X)}(\mathcal{F}, \mathcal{S}) \to \operatorname{Ext}^{i-1}_{\operatorname{Mod}(Y)}(\mathcal{F}, \mathcal{S}(Y)) \to \operatorname{Ext}^{i+1}_{\operatorname{Mod}(Y)}(\mathcal{F}, \mathcal{S}) \to$$

PROOF. These are standard excercises in homological algebra. The following elegant proof for 2. was suggested by the referee.

There is a change of rings spectral sequence

$$\mathrm{Ext}^p_Y(\mathcal{F}, \mathcal{E}xt^q_X(o_Y, \mathcal{S})) \Rightarrow \mathrm{Ext}^{p+q}_X(\mathcal{F}, \mathcal{S}) \tag{5.1}$$

(for this we only need that Y is (weakly) closed in X).

From the resolution

$$0 \to o_X(-Y) \to o_X \to o_Y \to 0$$

we find that

$$\mathcal{E}xt^i_X(o_Y, \mathcal{S}) = \begin{cases} \mathcal{S} & \text{if } i = 0 \\ \mathcal{S}(Y) & \text{if } i = 1 \\ 0 & \text{otherwise} \end{cases}$$

It follows that the spectral sequence (5.1) has only two rows and hence it degenerates into a long exact sequence which is precisely 2. □

Since $\mathcal{N}_{Y/X}$ is an invertible bimodule on Y it follows from [**6**, Prop. 6.8] that we have $\mathcal{N}_{Y/X} = \mathcal{N}_\tau$ for some line bundle $\mathcal{N}$ on Y and an automorphism τ of Y. Recall that by definition

$$(- \otimes_{o_Y} \mathcal{N}_\tau) = \tau_*(- \otimes_{\mathcal{O}_Y} \mathcal{N}) \tag{5.2}$$

By $\mathcal{C}_f$ we denote the finite length objects in $\mathrm{Mod}(X)$ whose Jordan-Holder quotients lie in $\mathrm{Mod}(Y)$. By $\mathcal{C}$ we denote the corresponding locally finite subcategory of $\mathrm{Mod}(X)$.

If $p \in Y$ then we denote by $\mathcal{C}_{f,p}$ the full subcategory of $\mathcal{C}_f$ consisting of objects whose Jordan-Holder quotients are among $(O_{\tau^n p})_n$. Again $\mathcal{C}_p$ is the corresponding locally closed subcategory of $\mathcal{C}$. From Propositions 5.1.2.2 it follows that $\mathcal{C}_f = \oplus_{p \in Y/\langle\tau\rangle} \mathcal{C}_{f,p}$, $\mathcal{C} = \oplus_{p\in Y/\langle\tau\rangle} \mathcal{C}_p$.

From the fact that $o_X(Y)/o_X = \mathcal{N}_\tau$ we deduce that

$$O_q(Y) \cong O_{\tau q} \tag{5.3}$$

In particular $\mathcal{C}_p$ is stable under $- \otimes_{o_X} o_X(Y)$. From this one deduces

PROPOSITION 5.1.3. *([**36**, Prop. 8.4]) $\mathcal{C}_p$ is closed under injective hulls in the category* $\mathrm{Mod}(X)$.

We now translate (and slightly generalize) the main result of [**36**] to our situation.

THEOREM 5.1.4. *We have the following.*

1. *There is a category equivalence $\hat{(-)}_p$ between $\mathcal{C}_p$ and the category* $\mathrm{Dis}(C_p)$ *(§4) for a certain pseudo-compact ring C_p. This ring C_p has the following form :*
   (a) *If $|O_\tau(p)| = \infty$ then C_p is given by the $\mathbb{Z}\times\mathbb{Z}$ lower triangular matrices with entries in $\hat{\mathcal{O}}_{Y,p}$. In this case p is regular on Y and thus we have $\hat{\mathcal{O}}_{Y,p} \cong k[[x]]$.*

(b) *If $|O_\tau(p)| = n$ then C_p is given by a ring of $n \times n$ matrices of the form*

$$\begin{pmatrix} R & RU & \cdots & RU \\ \vdots & \ddots & \ddots & \vdots \\ \vdots & & \ddots & RU \\ R & \cdots & \cdots & R \end{pmatrix}$$

where R is a complete local ring of the form

$$R = k\langle\langle x, y\rangle\rangle/(\psi)$$

with

$$\psi = yx - qxy + \text{higher terms} \tag{5.4}$$

for some $q \in k^*$, or

$$\psi = yx - xy - x^2 + \text{higher terms} \tag{5.5}$$

U is a regular normalizing element in $\operatorname{rad}(R)$ *such that $R/(U) = \hat{\mathcal{O}}_{Y,p}$.*

If p is not fixed under τ then p is regular on Y and also $U \notin \operatorname{rad}^2(R)$.

In all cases R carries the usual topology and C_p carries the corresponding product topology.

2. *Let $I = \mathbb{Z}$ if $|O_\tau(p)| = \infty$ and $I = \mathbb{Z}/n\mathbb{Z}$ if $|O_\tau(p)| = n$. In this way the elements of C_p correspond to $I \times I$-matrices. For $i \in I$ let e_i be the corresponding diagonal idempotent. Put $S_i = e_i C_p / \operatorname{rad}(e_i C_p)$. Then $(\mathcal{O}_{\tau^i p})^\wedge_p = S_i$.*
3. *Define the following normal element N of C_p.*
   (a) *If $|O_\tau(p)| = \infty$ then N is given by the matrix whose entries are everywhere zero except on the lower subdiagonal where they are one.*
   (b) *If $|O_\tau(p)| < \infty$ then*

$$N = \begin{pmatrix} 0 & \cdots & 0 & U \\ 1 & \ddots & & 0 \\ \vdots & \ddots & \ddots & \vdots \\ 0 & \cdots & 1 & 0 \end{pmatrix}$$

Let $\phi = N \cdot N^{-1}$ Then we have the following commutative diagram

$$\begin{array}{ccc} \mathcal{C}_p & \xrightarrow{-\otimes o_X(-Y)} & \mathcal{C}_p \\ \downarrow{\scriptstyle (\hat{-})_p} & & \downarrow{\scriptstyle (\hat{-})_p} \\ \operatorname{Dis}(C_p) & \xrightarrow{(-)_\phi} & \operatorname{Dis}(C_p) \end{array} \tag{5.6}$$

4. *If $\mathcal{F}$ is an object in $\mathcal{C}_p$ then one has the following commutative diagram.*

$$\begin{array}{ccc} (\mathcal{F}(-Y))^\wedge_p & \longrightarrow & (\mathcal{F})^\wedge_p \\ \cong\downarrow & & \| \\ ((\mathcal{F})^\wedge_p)_\phi & \xrightarrow{\cdot N} & (\mathcal{F})^\wedge_p \end{array}$$

where the top arrow is obtained from the inclusion $o_X(-Y) \hookrightarrow o_X$ and the left arrow from (5.6).

5. *Let* $\mathcal{F}$ *be a finite length object in* $\mathrm{Mod}(Y)$. *Then* $\hat{\mathcal{F}}_p$ *is a module over* $C_p/(N) = \prod_{q\in O_\tau(p)} \hat{O}_{Y,p}$ *and*

$$\hat{\mathcal{F}}_p = \prod_{q\in O_\tau(p)} \hat{\mathcal{F}}_{Y,q} \tag{5.7}$$

where we have written $\hat{(-)}_{Y,q}$ *for the ordinary completion at* q *on* Y.

In [**36**] this theorem was proved with $\mathcal{C}_p$ replaced by $\mathcal{C}_{f,p}$. However it is easy to see that one obtains the current theorem by taking direct limits.

Below we write m for the maximal ideal of R and m_i will be the maximal ideal of C_p corresponding to S_i. We also use S_i for the bimodule C_p/m_i.

R is clearly noetherian. C_p is noetherian if the orbit of p is finite, and locally noetherian otherwise. Furthermore it is also clear that C_p is cofinite.

The following was proved in [**36**, Thm 1.1.4].

LEMMA 5.1.5. *Every finite dimensional* C_p*-representation* F *is in* $\mathrm{PCFin}(C_p)$. *Hence* $F = \prod_i Fe_i$.

The ring C_p has another good property, which wasn't stated in [**36**].

PROPOSITION 5.1.6. *The ring* C_p *is coherent.*

PROOF. This is clear if the orbit of p is finite, so we assume it to be infinite.

We first show that if B is a finitely generated N-torsion free right pseudo-compact C_p-module then B is finitely presented.

We have an exact sequence of pseudo-compact modules

$$0 \to K \to C_p^m \to B \to 0$$

Tensoring with $C_{Y,p} \overset{\text{def}}{=} C_p/C_pN$ yields an exact sequence

$$0 \to K/NK \to C_{Y,p}^m \to B/NB \to 0$$

and using Lemma 4.9 we see that it is sufficient to show that K/NK is finitely generated. Now $C_{Y,p} = \prod_i \hat{O}_{Y,\tau^i p}$, and from the theory of discrete valuation rings we see that the number of generators of a submodule of $C_{Y,p}^m e_i = \hat{O}_{Y,\tau^i p}^m$ is bounded by m. This easily implies what we want.

Now we prove that C_p is coherent. We have to show that the kernel of an arbitrary map $\alpha : C_p^m \to C_p$ is finitely generated. Clearly $B = \operatorname{im}\alpha$ is pseudo-compact and finitely generated. We can now apply the result in the previous paragraph. □

We will also need :

PROPOSITION 5.1.7. *The global dimension of* C_p *is equal to two.*

PROOF. According to Corollaries 4.11 and 4.7 it suffices to show that the projective dimension of each S_i is equal to two.

Put $P_i = e_iC_p$. One easily checks that the minimal resolution of S_i is given by

$$0 \to P_{i-1} \to P_{i-1} \oplus P_i \to P_i \to S_i \to 0$$

This implies what we want. □

The following result will be used.

LEMMA 5.1.8. *Assume that p is a fixed point for τ. Then the multiplicity of p on Y is equal to the largest integer n such that $RU \subset \operatorname{rad}^n R$.*

PROOF. Put $S = R/(U)$ and let m be the maximal ideal of R. By Theorem 5.1.4.5 $S = \hat{\mathcal{O}}_{Y,p}$. Assume that $U \in m^{\mu} - m^{\mu+1}$. Equip R and S with the m-adic filtration. Since $\operatorname{gr} R$ is a domain (direct verification) it is easy to see that there is an exact sequence

$$0 \to \operatorname{gr} R(-\mu) \xrightarrow{\cdot U} \operatorname{gr} R \to \operatorname{gr} S \to 0 \tag{5.8}$$

Then from (5.8) we find that μ is equal to $\dim_k \operatorname{rad}^u S / \operatorname{rad}^{u+1} S$ for large u. Hence μ is the multiplicity of p in Y. □

5.2. Some computations

Our view point is that C_p encodes the local structure around a point $p \in Y$. In order to blow up p we will consequently need some computations in C_p. Our aim here is to prove Corollary 5.2.4 below. This corollary is easy if p is a fixed point for τ and fairly easy if p has infinite τ-orbit. So the main purpose will be to treat the case where $n = |O_\tau(p)|$ satisfies $2 \le n < \infty$. However we will develop a formalism which also includes the case $n = \infty$. Perhaps this has some independent interest. Our main result will be Proposition 5.2.2 which is however more elaborate than what we need for the application to Corollary 5.2.4.

Notations will be as in Theorem 5.1.4. m will be the maximal ideal of R and m_i will be the twosided maximal ideal of C_p corresponding to S_i. Note that usually i is a taken modulo n here.

DEFINITION 5.2.1. An array $a = (a_q)_{q\in\mathbb{Z}}$ with entries in $\mathbb{N} \cup \{+\infty\}$ will be called *good* if it is non-decreasing, bounded below, and if a_q is infinite for $t \gg 0$ and finite for $t \ll 0$. If a, b are good then $a \ge b$ iff $a_q \ge b_q$ for all q.

Our aim is to use good arrays as a bookkeeping device in order to study certain right C_p-modules in the case that $n \ge 2$. Let a be a good array and let I be as in Theorem 5.1.4. We distinguish two cases.

- **The case $n < \infty$.**
  Fix an arbitrary $x \in m - m^2$ such that $x - U \notin m^2$ for R. For $s \in \mathbb{Z}$ define

  $$H_{a,s} = \sum_{t\in\mathbb{Z}} x^{a_{s-nt}} U^t R$$

  where this sum is taken inside the fraction field of R and where by convention $x^\infty = 0$. The $H_{a,s}$ are clearly fractional right R ideals.
  Put

  $$P_a = (H_{a,0}, \dots, H_{a,n-1})$$

- **The case $n = \infty$.** We consider this as a limiting case of the previous case. We define P_a as the row matrices $(P_{a,j})_j$ with

  $$P_{a,j} = m^{a_j}$$

PROPOSITION 5.2.2. *In this proposition a, b, c will be good arrays. As above let $n = |I| \geq 2$. If $i \in I$ then*

$$\tilde{\imath} = \begin{cases} \text{the unique element of } \{0, \ldots, n-1\} \text{ congruent to } i & \text{if } n < \infty \\ i & \text{if } n = \infty \end{cases}$$

1. *The P_a are right C_p-modules (with the obvious C_p-action).*
2. *Let K be the fraction field of R and consider the P_a as submodules of K^I. Then*

$$P_a \subset P_b \iff a \geq b$$

3. *One also has*

$$P_a + P_b = P_{\inf(a,b)}$$
$$P_a \cap P_b = P_{\sup(a,b)}$$

4. *If $b \geq a$ and $a_i = \infty \iff b_i = \infty$ then P_a/P_b has finite length and the composition factors are given by*

$$\oplus_{i \in \mathbb{Z}} S_{\tilde{\imath}}^{b_i - a_i}$$

(with multiplicity).

5. *One has*

$$P_a m_l = P_c$$

where

$$c_q = \begin{cases} a_q & \text{if } q \ncong l \bmod n \\ \min(a_{q+1}, a_q + 1) & \text{if } q \cong l \bmod n \end{cases}$$

6. *One has*

$$\operatorname{rad} P_a = P_c$$

where

$$c_q = \begin{cases} a_q + 1 & \text{if } a_q \neq a_{q+1} \\ a_q & \text{otherwhise} \end{cases}$$

7. *One has*

$$P_a / \operatorname{rad} P_a = \bigoplus_{q \in \mathbb{Z}, a_q \neq a_{q+1}} S_{\tilde{q}}$$

8. *For $i \in I$ write $P_i = e_i C_p$. Then*

$$P_i = P_c$$

where

$$c = (\ldots, 0, \ldots, 0, \infty, \ldots)$$

with the first ∞ occuring in position $\tilde{\imath} + 1$.

PROOF. All this is fairly easy if $n = \infty$, so we concentrate on the case $n < \infty$.

It is easy to see that $x^i U^j$ is a topological k-basis for R and from the fact that U is normalizing one obtains the following alternative form of $H_{a,s}$.

$$H_{a,s} = \prod_{t \in \mathbb{Z}} \prod_{i \geq a_{s-nt}} k x^i U^t \tag{5.9}$$

From the fact that a_q is ascending we obtain $H_{a,s} \subset H_{a,s-1}$. We also have $H_{a,s-n} = H_{a,s}U^{-1}$. This easily implies 1.

If a, b are good then (5.9) implies that

$$\forall s : H_{a,s} \subset H_{b,s} \iff a \geq b \tag{5.10}$$

and also

$$H_{a,s} + H_{b,s} = H_{\inf(a,b),s} \tag{5.11}$$

$$H_{a,s} \cap H_{b,s} = H_{\sup(a,b),s} \tag{5.12}$$

It is clear that (5.10)(5.11) and (5.12) imply the corresponding properties for P_a, P_b. This proves 2. and 3.

If a, a' are such that

$$a'_q = \begin{cases} a_q & \text{if } q \neq i \\ a_q + 1 & \text{if } q = i,\ a_i \neq \infty \end{cases}$$

then it follows from (5.9) that

$$H_{a,s}/H_{a',s} = \begin{cases} R/m & \text{if } s \cong i \bmod n \\ 0 & \text{otherwise} \end{cases}$$

We deduce that

$$P_a/P_{a'} = S_{\bar{\imath}}$$

This yields 4.

We now compute $P_a m_l$ for an arbitrary $l \in I$. We find

$$(P_a m_l)_j = \begin{cases} H_{a,j} & \text{if } j \neq l \\ H_{a,j}m + H_{a,j+1} & \text{if } j = l \end{cases} \tag{5.13}$$

To compute the righthand side of (5.13) we use the observation that $U^{-b}xU^b \not\cong U \bmod m^2$ and hence $m = UR + U^{-b}xU^bR$. Thus

$$x^aU^bm = x^aU^b(UR + U^{-b}xU^bR) = x^{a+1}U^bR + x^aU^{b+1}R$$

One now easily obtains 5. Items 6. and 7. are consequences of the fact that $\operatorname{rad} P_a = \bigcap_l P_a m_l$. Finally 8. is a simple verification which we leave to the reader. □

EXAMPLE 5.2.3. Let $n = \infty$, $a = (\dots, 0, 0, 1, 1, 1, 3, \infty, \infty, \dots)$ where the first ∞ occurs in position 2. Then

$$P_a/\operatorname{rad} P_a = S_1 \oplus S_0 \oplus S_{-3} \tag{5.14}$$

If $n = 3$ and a is the same then

$$P_a/\operatorname{rad} P_a = S_{\bar{0}}^2 \oplus S_{\bar{1}} \tag{5.15}$$

Note that going from (5.14) to (5.15) amounts to introducing periodicity modulo 3 among the S_i.

COROLLARY 5.2.4. $\dim_k(C_p/m_0m_{-1}m_{-2}\cdots m_{-p+1}) = \frac{p(p+1)}{2}$

PROOF. This is clear if $n = 1$. In the case $n \geq 2$ we use the fact that we have $C_p = \prod_{i \in I} P_i$. The corollary now follows easily from 5. and 8. of the foregoing proposition. □

5.3. Completion of objects in $\mathrm{mod}(X)$

We have already defined $\hat{(-)}_p$ on $\mathcal{C}_p$. In a different direction, it is also possible to extend $\hat{(-)}_p$ to $\mathrm{mod}(X)$, but then it loses some of its good properties. For $\mathcal{F} \in \mathrm{mod}(X)$ define

$$\hat{\mathcal{F}}_p = \varprojlim_{\mathcal{F}/\mathcal{F}' \in \mathcal{C}_p} (\mathcal{F}/\mathcal{F}')_p^\wedge$$

THEOREM 5.3.1. *$(-)_p^\wedge$ is an exact functor from $\mathrm{mod}(X)$ to $\mathrm{PC}(C_p)$. The analogs of 3.,4.,5. of Theorem 5.1.4 hold for $\mathcal{F} \in \mathrm{mod}(X)$. If $\mathcal{F} \in \mathrm{mod}(X)$ then $\hat{\mathcal{F}}_p$ is finitely presented. Furthermore, $\hat{\mathcal{F}}_p$ lies in $\mathrm{pc}(C_p)$ if and only if the intersection of the support of $\mathcal{F}/\mathcal{F}(-Y) \in \mathrm{Mod}(Y)$ and the orbit of p is finite.*

PROOF. Let

$$0 \to \mathcal{F} \xrightarrow{\phi} \mathcal{G} \xrightarrow{\theta} \mathcal{H} \to 0$$

be an exact sequence in $\mathrm{mod}(X)$. We have to show that

$$0 \to \varprojlim_{\mathcal{F}/\mathcal{F}' \in \mathcal{C}_p} (\mathcal{F}/\mathcal{F}')_p^\wedge \to \varprojlim_{\mathcal{G}/\mathcal{G}' \in \mathcal{C}_p} (\mathcal{G}/\mathcal{G}')_p^\wedge \to \varprojlim_{\mathcal{H}/\mathcal{H}' \in \mathcal{C}_p} (\mathcal{H}/\mathcal{H}')_p^\wedge \to 0$$

is exact. Given the exactness of $(-)_p^\wedge$ on finite length objects and the exactness of $\varprojlim$ on pseudo-compact modules this means we have to show that

1. $\theta(\mathcal{G}')_{\mathcal{G}/\mathcal{G}' \in \mathcal{C}_p}$ is cofinal in $(\mathcal{H}')_{\mathcal{H}/\mathcal{H}' \in \mathcal{C}_p}$
2. $\phi^{-1}(\mathcal{G}')_{\mathcal{G}/\mathcal{G}' \in \mathcal{C}_p}$ is cofinal in $(\mathcal{F}')_{\mathcal{F}/\mathcal{F}' \in \mathcal{C}_p}$.

The first statement is clear. The second statement is the Artin-Rees condition, which is equivalent to $\mathcal{C}_p$ being closed under injective hulls in $\mathrm{Mod}(X)$. This is precisely Proposition 5.1.3. The fact that the analog of Theorem 5.1.4 holds is easy to see.

Now we prove that $\hat{\mathcal{F}}_p$ is finitely presented. According to lemma 4.18 it is sufficient to show that for every i and every $q \in O_\tau(p)$, $\mathrm{Ext}^i_{\mathrm{PC}(C_p)}(\hat{\mathcal{F}}_p, (\mathcal{O}_q)_p^\wedge)$ has finite dimension, bounded independently of q. By Proposition 5.3.4 below we have

$$\mathrm{Ext}^i_{\mathrm{PC}(C_p)}(\hat{\mathcal{F}}_p, (\mathcal{O}_q)_p^\wedge)) = \mathrm{Ext}^i_{\mathrm{Mod}(X)}(\mathcal{F}, \mathcal{O}_q)$$

and the dimension of the righthand side of this equation is indeed bounded independently of q by lemma 5.3.2.

Now we concentrate on the second part of the theorem. We have an exact sequence in $\mathrm{Mod}(X)$

$$\mathcal{F}(-Y) \to \mathcal{F} \to \mathcal{F}/\mathcal{F}(-Y) \to 0$$

which yields by the analog of Theorem 5.1.4.5 an exact sequence in $\mathrm{PC}(C_p)$

$$(\hat{\mathcal{F}}_p)_\phi \xrightarrow{\cdot N} \hat{\mathcal{F}}_p \to \prod_{q \in O_\tau(p)} (\mathcal{F}/\mathcal{F}(-Y))_{Y,q}^\wedge \to 0$$

By [**36**, Prop 3.22] $\hat{\mathcal{F}}_p$ will be noetherian if and only if $\prod_{q \in O_\tau(p)} (\mathcal{F}/\mathcal{F}(-Y))_{Y,q}^\wedge$ is noetherian, that is, if and only if, for almost all $q \in O_\tau(p)$ we have $(\mathcal{F}/\mathcal{F}(-Y))_{Y,q}^\wedge = 0$. This is equivalent to the intersection of the support of $\mathcal{F}/\mathcal{F}(-Y)$ with the τ-orbit of p being finite. □

We now supply the details that were used in the proof of the previous theorem.

LEMMA 5.3.2. *Assume that* $\mathcal{F} \in \mathrm{mod}(X)$, $q \in Y$. *Then* $\dim \mathrm{Ext}^i_{\mathrm{Mod}(X)}(\mathcal{F}, \mathcal{O}_q)$ *is finite, and bounded independently of* q.

PROOF. As usual, one reduces to one of the following cases.

1. The canonical map $\mathcal{F}(-Y) \to \mathcal{F}$ is monic.
2. $\mathcal{F}$ is in $\mathrm{mod}(Y)$.

Using Proposition 5.1.2, we can then reduce to showing that if $\mathcal{G} \in \mathrm{mod}(Y)$ then $\dim \mathrm{Ext}^i_{\mathrm{Mod}(Y)}(\mathcal{G}, \mathcal{O}_q)$ is finite and bounded independently of q. This now follows from the fact that $\mathcal{G}$ has a resolution consisting of vector bundles around Y. □

The following result is useful.

LEMMA 5.3.3. $\hat{\mathcal{O}}_{X,p} \cong C_p$ *as right* C_p*-modules.*

PROOF. The analogous result for Y is trivially true. We can lift this to X using Nakayama's lemma (lemma 4.8). □

We now have to be able to compare Ext-groups in $\mathrm{PC}(C_p)$ and $\mathrm{Mod}(X)$. In fact we have the following result.

PROPOSITION 5.3.4. *Assume that* $\mathcal{F} \in \mathrm{mod}(X)$ *and* $\mathcal{S} \in \mathcal{C}_p$. *Then*

$$\mathrm{Ext}^i_{\mathrm{Top}(C_p)}(\hat{\mathcal{F}}_p, \hat{\mathcal{S}}_p) = \mathrm{Ext}^i_{\mathrm{Mod}(X)}(\mathcal{F}, \mathcal{S})$$

PROOF. We can reduce to the case $i = 0$ by replacing $\mathcal{S}$ with an injective resolution in $\mathcal{C}_p$. We have

$$\begin{aligned}
\mathrm{Hom}_{\mathrm{Top}(A)}(\hat{\mathcal{F}}_p, \hat{\mathcal{S}}_p) &= \mathrm{Hom}_{\mathrm{Top}(A)}(\varprojlim_{\mathcal{F}'} (\mathcal{F}/\mathcal{F}')^\wedge_p, \hat{\mathcal{S}}_p) \\
&= \varinjlim_{\mathcal{F}'} \mathrm{Hom}_{\mathrm{Dis}(A)}((\mathcal{F}/\mathcal{F}')^\wedge_p, \hat{\mathcal{S}}_p) \qquad \text{(Proposition 4.3)} \\
&= \varinjlim_{\mathcal{F}'} \mathrm{Hom}_{\mathrm{Mod}(X)}(\mathcal{F}/\mathcal{F}', \mathcal{S}) \\
&= \mathrm{Hom}_{\mathrm{Mod}(X)}(\mathcal{F}, \mathcal{S})
\end{aligned}$$

Here $\mathcal{F}'$ runs through the subobjects in $\mathcal{F}$ such that $\mathcal{F}/\mathcal{F}' \in \mathcal{C}_{f,p}$. □

COROLLARY 5.3.5. *Assume* $\mathcal{F} \in \mathrm{mod}(X)$. *Then* $\hat{(-)}_p$ *defines a one-one correspondence between open subobjects of* $\hat{\mathcal{F}}_p$ *and the subobjects* $\mathcal{F}'$ *of* $\mathcal{F}$ *such that* $\mathcal{F}/\mathcal{F}' \in \mathcal{C}_p$.

PROOF. Assume for example that $H \subset \hat{\mathcal{F}}_p$ is open. Put $S = \hat{\mathcal{F}}_p/H$. Then $S \in \mathrm{PCFin}(C_p)$. By Theorem 5.1.4 we have $S = \hat{\mathcal{S}}_p$ for some $\mathcal{S} \in \mathcal{C}_{f,p}$. Let $q : \hat{\mathcal{F}}_p \to S$ be the quotient map. According to Proposition 5.3.4, q corresponds to a map $p : \mathcal{F} \to \mathcal{S}$. Define $\mathcal{H} = \ker p$. Then $\hat{\mathcal{H}}_p = H$. □

5.4. Completion of bimodules

If is also possible to define $\hat{(-)}_p$ on certain $o_X - o_X$-bimodules. Unless otherwise specified, when we say "bimodule" we mean an object of $\mathrm{Bimod}(o_X - o_X)$.

DEFINITION 5.4.1. We say that an $o_X - o_X$ bimodule $\mathcal{M}$ is $\mathcal{C}_p$-preserving if both $- \otimes_{o_X} \mathcal{M}$ and $\mathcal{H}om_{o_X}(\mathcal{M}, -)$ preserve $\mathcal{C}_p$.

In this section we will write $(-)^{\sim}_{p} : \mathrm{Dis}(C_p) \to \mathcal{C}_p$ for the inverse of the functor $(-)^{\wedge}_{p}$.

Assume that $\mathcal{M}$ is a coherent $\mathcal{C}_p$ preserving bimodule. Then $-\otimes_{o_X}\mathcal{M}$ preserves $\mathcal{C}_{f,p}$ by 3.1.6. We define

$$\hat{\mathcal{M}}_p = \varprojlim_I ((C_p/I)^{\sim}_{p} \otimes_{o_X} \mathcal{M})^{\wedge}_{p} \tag{5.16}$$

where I runs over the open twosided ideals in C_p.

PROPOSITION 5.4.2. *Assume that $\mathcal{M}$ is a coherent $\mathcal{C}_p$ preserving $o_X - o_X$-bimodule. Then $\hat{\mathcal{M}}_p$ is a bipseudo-compact C_p-bimodule.*

PROOF. If follows from functoriality that $((C_p/I)^{\sim}_{p} \otimes_{o_X} \mathcal{M})^{\wedge}_{p}$ is annihilated by I on the left. On the right it is in $\mathrm{PCFin}(C_p)$ and hence it is also annihilated by some open twosided ideal $J \subset C_p$. Thus $((C_p/I)^{\sim}_{p} \otimes_{o_X} \mathcal{M})^{\wedge}_{p}$ is bipseudo-compact, and hence so is the inverse limit. □

We have the following analog of Proposition 5.3.4.

PROPOSITION 5.4.3. *Assume that $\mathcal{M}$ is a coherent $\mathcal{C}_p$-preserving bimodule and $\mathcal{S} \in \mathcal{C}_p$. Then $\mathrm{Ext}^i_{\mathrm{Top}(C_p)}(\hat{\mathcal{M}}_p, \hat{\mathcal{S}}_p) \in \mathrm{Dis}(C_p)$ and $\mathcal{E}xt^i_{\mathrm{Mod}(X)}(\mathcal{M}, \mathcal{S}) \in \mathcal{C}_p$. Furthermore as objects of $\mathrm{Dis}(C_p)$*

$$\mathrm{Ext}^i_{\mathrm{Top}(C_p)}(\hat{\mathcal{M}}_p, \hat{\mathcal{S}}_p) = \mathcal{E}xt^i_{\mathrm{Mod}(X)}(\mathcal{M}, \mathcal{S})^{\wedge}_{p}$$

PROOF. That $\mathcal{E}xt^i_{\mathrm{Mod}(X)}(\mathcal{M}, \mathcal{S}) \in \mathcal{C}_p$ follows from the fact that $\mathcal{H}om(\mathcal{M}, -)$ is $\mathcal{C}_p$ preserving and Proposition 5.1.3. The fact that $\mathrm{Ext}^i_{\mathrm{Top}(C_p)}(\hat{\mathcal{M}}_p, \hat{\mathcal{S}}_p) \in \mathrm{Dis}(C_p)$ is precisely Proposition 4.19.

Below $\mathcal{U}$ will run through $(C_p/I)^{\sim}_{p}$, where I is an open twosided ideal in C_p. We have

(5.17)
$$\begin{aligned}
\mathrm{Ext}^i_{\mathrm{Top}(C_p)}(\hat{\mathcal{M}}_p, \hat{\mathcal{S}}_p) &= \mathrm{Ext}^i_{\mathrm{Top}(C_p)}(\varprojlim_{\mathcal{U}} (\mathcal{U} \otimes_{o_X} \mathcal{M})^{\wedge}_{p}, \hat{\mathcal{S}}_p)\\
&= \varinjlim_{\mathcal{U}} \mathrm{Ext}^i_{\mathrm{Dis}(C_p)}((\mathcal{U} \otimes_{o_X} \mathcal{M})^{\wedge}_{p}, \hat{\mathcal{S}}_p) \qquad \text{(Proposition 4.3)}\\
&= \varinjlim_{\mathcal{U}} \mathrm{Ext}^i_{\mathcal{C}_p}(\mathcal{U} \otimes_{o_X} \mathcal{M}, \mathcal{S}) \qquad \text{(Theorem 5.1.4)}\\
&= \varinjlim_{\mathcal{U}} \mathrm{Ext}^i_{\mathrm{Mod}(X)}(\mathcal{U} \otimes_{o_X} \mathcal{M}, \mathcal{S}) \quad \text{(Proposition 5.1.3)}
\end{aligned}$$

Let $E^{\cdot}$ be an injective resolution of $\mathcal{S}$ in $\mathcal{C}_p$. Then

$$\begin{aligned}
\mathrm{Ext}^i_{\mathrm{Mod}(X)}(\mathcal{U} \otimes_{o_X} \mathcal{M}, \mathcal{S}) &= H^i(\mathrm{Hom}_{\mathrm{Mod}(X)}(\mathcal{U} \otimes_{o_X} \mathcal{M}, E^{\cdot}))\\
&= H^i(\mathrm{Hom}_{\mathrm{Mod}(X)}(\mathcal{U}, \mathcal{H}om_{\mathrm{Mod}(X)}(\mathcal{M}, E^{\cdot})))\\
&= H^i(\mathrm{Hom}_{\mathrm{Dis}(C_p)}(\hat{\mathcal{U}}_p, \mathcal{H}om_{\mathrm{Mod}(X)}(\mathcal{M}, E^{\cdot})^{\wedge}_{p}))\\
&= H^i(\mathrm{Hom}_{\mathrm{Dis}(C_p)}(C_p/I, \mathcal{H}om_{\mathrm{Mod}(X)}(\mathcal{M}, E^{\cdot})^{\wedge}_{p}))
\end{aligned}$$

Combining this with (5.17) yields

$$\begin{aligned}\mathrm{Ext}^i_{\mathrm{Top}(C_p)}(\hat{\mathcal{M}}_p,\hat{\mathcal{S}}_p) &= \varinjlim_I H^i(\mathrm{Hom}_{\mathrm{Dis}(C_p)}(C_p/I,\mathcal{H}om_{\mathrm{Mod}(X)}(\mathcal{M},E^\cdot)^\wedge_p)\\ &= H^i(\mathrm{Hom}_{\mathrm{Top}(C_p)}(\varprojlim_I(C_p/I),\mathcal{H}om_{\mathrm{Mod}(X)}(\mathcal{M},E^\cdot)^\wedge_p))\\ &= H^i(\mathrm{Hom}_{\mathrm{Top}(C_p)}(C_p,\mathcal{H}om_{\mathrm{Mod}(X)}(\mathcal{M},E^\cdot)^\wedge_p))\\ &= \mathcal{E}xt^i_{\mathrm{Mod}(X)}(\mathcal{M},\mathcal{S})^\wedge_p \quad \square\end{aligned}$$

PROPOSITION 5.4.4. *The functor $(-)^\wedge_p$ preserves short exact sequences of coherent $\mathcal{C}_p$-preserving bimodules.*

PROOF. Let

$$0\to\mathcal{K}\to\mathcal{M}\to\mathcal{N}\to 0$$

be an exact sequence of coherent $\mathcal{C}_p$ preserving bimodules. We have to show that

$$0\to\hat{\mathcal{K}}_p\to\hat{\mathcal{M}}_p\to\hat{\mathcal{N}}_p\to 0 \tag{5.18}$$

is exact. Since the bimodules in (5.18) are in $\mathrm{PC}(C_p)$ (Proposition 5.4.2) it suffices to show that

(5.19)

$$0\to\mathrm{Hom}_{\mathrm{Top}(C_p)}(\hat{\mathcal{N}}_p,\hat{E}_p)\to\mathrm{Hom}_{\mathrm{Top}(C_p)}(\hat{\mathcal{M}}_p,\hat{E}_p)\to\mathrm{Hom}_{\mathrm{Top}(C_p)}(\hat{\mathcal{K}}_p,\hat{E}_p)\to 0$$

is exact where E is the sum of the injective hulls of the $\mathcal{O}_{\tau^i p}$ (one uses the fact that $\mathrm{Hom}_{\mathrm{Top}(C_p)}(-,\hat{E}_p)$ is exact and faithful on $\mathrm{PC}(C_p)$). Now by Proposition 5.4.3 it follows that (5.19) is obtained from completing the following exact sequence in $\mathcal{C}_p$.

$$0\to\mathcal{H}om_{\mathrm{Mod}(X)}(\mathcal{N},E)\to\mathcal{H}om_{\mathrm{Mod}(X)}(\mathcal{M},E)\to\mathcal{H}om_{\mathrm{Mod}(X)}(\mathcal{K},E)\to 0$$

and hence we are done! $\square$

PROPOSITION 5.4.5. *Completion commutes with tensor product in the following sense. Let $\mathcal{F}$ be a coherent object in* Mod(X) *and let $\mathcal{M}$, $\mathcal{N}$ be coherent $\mathcal{C}_p$ preserving bimodules. Then there are natural isomorphisms*

1. $(\mathcal{F}\otimes_{\mathcal{O}X}\mathcal{N})^\wedge_p=\hat{\mathcal{F}}_p\hat{\otimes}_{C_p}\hat{\mathcal{N}}_p$ *(note that by Theorem 5.3.1 and lemma 4.14 we may replace "$\hat{\otimes}$" by "$\otimes$").*
2. $(\mathcal{M}\otimes_{\mathcal{O}X}\mathcal{N})^\wedge_p=\hat{\mathcal{M}}_p\hat{\otimes}_{C_p}\hat{\mathcal{N}}_p$

PROOF. 1. By definition we have

$$(\mathcal{F}\otimes_{\mathcal{O}X}\mathcal{N})^\wedge_p=\varprojlim_{\mathcal{T}}((\mathcal{F}\otimes_{\mathcal{O}X}\mathcal{N})/\mathcal{T})^\wedge_p$$

where $\mathcal{T}$ runs through the subobjects of $\mathcal{F}\otimes_{\mathcal{O}X}\mathcal{N}$ such that $(\mathcal{F}\otimes_{\mathcal{O}X}\mathcal{N})\mathcal{T}\in\mathcal{C}_p$. Now we claim that for every $\mathcal{T}$ there exists a $\mathcal{F}'\subset\mathcal{F}$ such that $\mathcal{F}/\mathcal{F}'\in\mathcal{C}_p$ and such that the image of $\mathcal{F}'\otimes_{\mathcal{O}X}\mathcal{N}$ in $\mathcal{F}\otimes_{\mathcal{O}X}\mathcal{N}$ is contained in $\mathcal{T}$.

Let $\mathcal{Q}=(\mathcal{F}\otimes_{\mathcal{O}X}\mathcal{N})/\mathcal{T}$ and let $\mathcal{F}\otimes_{\mathcal{O}X}\mathcal{N}\to\mathcal{Q}$ be the quotient map. By adjointness there is a corresponding map $\mathcal{F}\to\mathcal{H}om(\mathcal{N},\mathcal{Q})$. We define $\mathcal{F}'$ as the kernel of this map. This $\mathcal{F}'$ has the properties we want.

We obtain

$$(\mathcal{F}\otimes_{\mathcal{O}X}\mathcal{N})^\wedge_p=\varprojlim_{\mathcal{F}'}((\mathcal{F}/\mathcal{F}')\otimes_{\mathcal{O}X}\mathcal{N})^\wedge_p \tag{5.20}$$

For $K \in \mathrm{PC}(C_p)$ we define the functor

$$F(K) = \varprojlim_{K'} ((K/K')_p^{\sim} \otimes_{o_X} \mathcal{N})_p^{\wedge}$$

where K' runs through the open subobjects of K. One shows that this functor is right exact and commutes with products. Furthermore, according to (5.16) we have $F(C_p) = \hat{\mathcal{N}}_p$. By (5.20) and Corollary 5.3.5

$$\begin{aligned} F(\hat{\mathcal{F}}_p) &= \varprojlim((\mathcal{F}/\mathcal{F}') \otimes_{o_X} \mathcal{N})_p^{\wedge} \\ &= (\mathcal{F} \otimes_{o_X} \mathcal{N})_p^{\wedge} \end{aligned}$$

Now by Theorem 5.3.1 we know that $\hat{\mathcal{F}}_p$ is coherent. Take a presentation

$$C_p^n \to C_p^m \to \hat{\mathcal{F}}_p \to 0$$

By right exactness of tensor product we find that the cokernel of $(C_p^n \to C_p^m) \otimes_{C_p} \hat{\mathcal{N}}_p$ is equal to $\hat{\mathcal{F}}_p \otimes_{o_X} \hat{\mathcal{N}}_p$. On the other hand by right exactness of F we find that this cokernel is equal to $F(\hat{\mathcal{F}}_p) = (\mathcal{F} \otimes \mathcal{N})_p^{\wedge}$.

2. This follows from 1. We have

$$\begin{aligned} (\mathcal{M} \otimes \mathcal{N})_p^{\wedge} &= \varprojlim_I ((C_p/I)_p^{\sim} \otimes_{o_X} \mathcal{M} \otimes_{o_X} \mathcal{N})_p^{\wedge} \\ &= \varprojlim_I ((C_p/I)_p^{\sim} \otimes_{o_X} \mathcal{M})_p^{\wedge} \hat{\otimes}_{C_p} \hat{\mathcal{N}}_p \\ &= \varprojlim_I ((C_p/I) \hat{\otimes}_{C_p} \hat{\mathcal{M}}_p \hat{\otimes}_{C_p} \hat{\mathcal{N}}_p \\ &= \hat{\mathcal{M}}_p \hat{\otimes}_{C_p} \hat{\mathcal{N}}_p \end{aligned}$$

where we have used the fact that $\hat{\otimes}$ commutes with inverse limits. □

5.5. The category $\tilde{\mathcal{C}}_{f,p}$

Let $q \in Y$ then we have closed embedding so quasi-schemes $q \subset Y \subset X$. Hence we we have a corresponding quotient $o_X \to o_q$.

LEMMA 5.5.1. 1. *Assume that* $\mathcal{F} \in \mathrm{Mod}(X)$. *Then there is a canonical isomorphism*

$$\mathcal{H}om_{\mathrm{Mod}(X)}(o_q, \mathcal{F}) \cong \mathrm{Hom}_{\mathrm{Mod}(X)}(\mathcal{O}_q, \mathcal{F}) \otimes_k \mathcal{O}_q \tag{5.21}$$

2. *If* $\mathcal{F} \in \mathrm{mod}(X)$ *then there is a canonical isomorphism*

$$\mathcal{F} \otimes_{o_X} o_q \cong \mathrm{Hom}_{\mathrm{Mod}(X)}(\mathcal{F}, \mathcal{O}_q)^* \otimes_k \mathcal{O}_q \tag{5.22}$$

3. *We have a canonical identification*

$$o_q \otimes_{o_X} o_X(Y) \cong o_X(Y) \otimes_{o_X} o_{\tau q} \quad \square \tag{5.23}$$

PROOF. 1. According to lemma 3.3.1.3 $\mathcal{H}om_{\mathrm{Mod}(X)}(o_q, \mathcal{F})$ picks out the maximal subobject of $\mathcal{F}$ which lies in $\mathrm{Mod}(q)$. It is easy to see that the functor $\mathrm{Hom}_{\mathrm{Mod}(X)}(\mathcal{O}_q, \mathcal{F}) \otimes_k \mathcal{O}_q$ does the same thing.

2. This is similar to 1., using lemma 3.3.1.5.

3. It suffices to show that the left and righthand side of (5.23) take the same values on arbitrary $\mathcal{F} \in \mathrm{Mod}(X)$.

   We have

$$\begin{aligned}\mathcal{H}om_{\mathrm{Mod}(X)}(o_q \otimes_{o_X} o_X(Y), \mathcal{F}) &= \mathcal{H}om_{\mathrm{Mod}(X)}(o_q, \mathcal{F}(-Y)) \\ &= \mathrm{Hom}(\mathcal{O}_q, \mathcal{F}(-Y)) \otimes_k \mathcal{O}_q \\ &= \mathrm{Hom}(\mathcal{O}_q(Y), \mathcal{F})) \otimes_k \mathcal{O}_q\end{aligned} \tag{5.24}$$

   and

$$\begin{aligned}\mathcal{H}om_{\mathrm{Mod}(X)}(o_X(Y) \otimes_{o_X} o_{\tau q}, \mathcal{F}) &= \mathcal{H}om_{\mathrm{Mod}(X)}(o_X(Y), \mathcal{H}om_{\mathrm{Mod}(X)}(o_{\tau q}, \mathcal{F})) \\ &= (\mathrm{Hom}_{\mathrm{Mod}(X)}(\mathcal{O}_{\tau q}, \mathcal{F}) \otimes_k \mathcal{O}_{\tau q})(-Y) \\ &= \mathrm{Hom}_{\mathrm{Mod}(X)}(\mathcal{O}_{\tau q}, \mathcal{F}) \otimes_k \mathcal{O}_{\tau q}(-Y)\end{aligned} \tag{5.25}$$

   Now by our conventions $\mathcal{O}_q(Y) \cong \mathcal{O}_{\tau q}$ (see (5.3)). From this it follows that the righthand sides in (5.24) and (5.25) are isomorphic. The reason that this is canonical is that we use the (non-canonical) identification $\mathcal{O}_q(Y) \cong \mathcal{O}_{\tau q}$ twice, in such a way that the ambiguities cancel. □

Let $\mathrm{cohBIMOD}(o_x - o_X)$ be the full subcategory of $\mathrm{BIMOD}(o_X - o_X)$ consisting of coherent objects. According to Corollary 3.1.8 this is an abelian subcategory of $\mathrm{BIMOD}(o_X - o_X)$, closed under extensions.

PROPOSITION 5.5.2. *o_q is a simple object in* $\mathrm{cohBIMOD}(o_X - o_X)$.

PROOF. An object in $\mathrm{cohBIMOD}(o_X - o_X)$ is determined by the values it takes on indecomposable injectives. We have by lemma 5.5.1.1

$$\mathcal{H}om_{\mathrm{Mod}(X)}(o_q, E) = \begin{cases} \mathcal{O}_q & \text{if } E \text{ is the injective hull of } \mathcal{O}_q \\ 0 & \text{otherwise} \end{cases}$$

Hence a proper left exact subfunctor of $\mathcal{H}om_{\mathrm{Mod}(X)}(o_q, -)$ commuting with direct sums will be the zero-functor. □

From this proposition we easily deduce

COROLLARY 5.5.3. *The objects $o_q \otimes_{o_X} o_X(nY)$ with $n \in \mathbb{Z}$ are simple objects in* $\mathrm{cohBIMOD}(o_X - o_X)$.

Now we define $\tilde{\mathcal{C}}_{f,p}$ as the full subcategory of $\mathrm{cohBIMOD}(o_X - o_X)$ consisting of finite length objects whose Jordan-Holder quotients are of the form $o_{\tau^m p} \otimes o_X(nY)$, $m, n \in \mathbb{Z}$.

PROPOSITION 5.5.4. *Assuming that $\mathcal{C}$, $\mathcal{D}$ are abelian categories possessing an injective cogenerator. Assume that we have an exact sequence in* $\mathrm{BIMOD}(\mathcal{C} - \mathcal{D})$

$$0 \to \mathcal{K} \to \mathcal{M} \to \mathcal{N} \to 0$$

with $\mathcal{N} \in \mathrm{Bimod}(\mathcal{C} - \mathcal{D})$, such that $R^1 \prod_{i \in I} \mathcal{H}om_{\mathcal{D}}(\mathcal{N}, E_i) = 0$ (see the discussion after Proposition 3.4.5) for all families of injectives $(E_i)_{i \in I}$ in $\mathcal{D}$. Then $\mathcal{K} \in \mathrm{Bimod}(\mathcal{C} - \mathcal{D})$ if and only if $\mathcal{M} \in \mathrm{Bimod}(\mathcal{C} - \mathcal{D})$.

PROOF. We have to show that $\mathcal{H}om_{\mathcal{D}}(\mathcal{K}, -)$ commutes products when evaluated on injectives, if and only if the same holds for $\mathcal{H}om_{\mathcal{D}}(\mathcal{M}, -)$. Using the hypotheses this is a direct consequence of the five-lemma. □

PROPOSITION 5.5.5. *Assume that* $\mathcal{S}$ *is an object in* $\tilde{\mathcal{C}}_{f,p}$. *Then the derived product* $R^1 \prod_{i\in I} \mathcal{H}om_{\mathrm{Mod}(X)}(\mathcal{S}, E_i)$ *is zero for all families of injectives* $(E_i)_{i\in I}$ *in* $\mathrm{Mod}(X)$.

PROOF. Using the long exact sequence for $R\prod_{i\in I}$ one sees that it is sufficient that $R^1 \prod_{i\in I} \mathcal{H}om_{\mathrm{Mod}(X)}(o_q, E_i) = 0$ for $q \in Y$ and for all families of injectives in $\mathrm{Mod}(X)$.

If $\mathcal{M} \in \mathrm{Mod}(X)$ then $\mathcal{H}om_{\mathrm{Mod}(X)}(o_q, \mathcal{M}) \in \mathrm{Mod}(q)$. By Definition 3.4.9 it now suffices to show that q is very well closed in X. Since q is obviously defined by an ideal in o_Y, we have that q is well closed in Y by Proposition 3.4.7. $\mathrm{Mod}(q)$ is of course equivalent to the category of k-vectorspaces. In particular $\mathrm{Mod}(q)$ has exact direct products and hence q is very well closed in Y by Corollary 3.4.11.

Furthermore Y is also defined by an ideal inside X and hence by Proposition 3.4.7 it follows that Y is well closed in X. Now we invoke Proposition 3.4.10 which yields that q is very well closed in X. □

COROLLARY 5.5.6. *If we have an exact sequence in* $\mathrm{BIMOD}(o_X - o_X)$

$$0 \to \mathcal{M} \to \mathcal{N} \to \mathcal{S} \to 0$$

with $\mathcal{S} \in \tilde{\mathcal{C}}_{f,p}$ *then* $\mathcal{M} \in \mathrm{Bimod}(o_X - o_X)$ *if and only if* $\mathcal{N} \in \mathrm{Bimod}(o_X - o_X)$.

PROPOSITION 5.5.7. $\tilde{\mathcal{C}}_{f,p} \subset \mathrm{Bimod}(o_X - o_X)$.

PROOF. Let $\mathcal{S} \in \tilde{\mathcal{C}}_{f,p}$. By induction there is a short exact sequence in $\tilde{\mathcal{C}}_{f,p}$

$$0 \to \mathcal{S}_1 \to \mathcal{S} \to \mathcal{S}_2 \to 0$$

such that $S_i \in \mathrm{Bimod}(o_X - o_X)$. It now suffices to apply Corollary 5.5.6 above. □

Let us denote by $\mathrm{PCFin}(C_p - C_p)$ the full subcategory of $\mathrm{PC}(C_p - C_p)$ consisting of finite length objects. Thus $\mathrm{PCFin}(C_p - C_p) = \mathrm{PCFin}(C_p^\circ \hat{\otimes} C_p)$. Recall that $C_p/\mathrm{rad}(C_p) = k^I$ where I is as in Theorem 5.1.4. Thus the twosided bipseudo-compact maximal ideals in C_p are naturally indexed by I. The corresponding simple modules were denoted by S_i in Theorem 5.1.4. Below we will view the S_i as C_p-bimodules.

It is clear that $(C_p)^\circ \hat{\otimes} C_p$ has a normalizing sequence given by $N \hat{\otimes} 1$ and $1 \hat{\otimes} N$ and the quotient is given by $(C_p/(N))^\circ \hat{\otimes} (C_p/(N))$ which is locally noetherian. It follows from [**36**, Prop. 3.23] that $(C_p)^\circ \hat{\otimes} C_p$ is locally noetherian. Hence in particular the category $\mathrm{PCFin}(C_p - C_p)$ is given by the category of C_p-bimodules which are finite extensions of the S_i.

The functor $\hat{(-)}_p$ is defined on $\tilde{\mathcal{C}}_{f,p}$. We have

LEMMA 5.5.8. $(o_{\tau^i p})_p^\wedge = S_i$

PROOF. We use (5.22). Let $q = \tau^i p$.

$$\begin{aligned}
(o_q)_p^\wedge &= \varprojlim_I ((C_p/I)_p^\sim \otimes_{C_p} o_q)_p^\wedge \\
&= \varprojlim_I ((\mathcal{O}_q)_p^\wedge \otimes_k \mathrm{Hom}_{\mathrm{Mod}(X)}((C_p/I)_p^\sim, \mathcal{O}_q)^*) \qquad \text{(lemma 5.5.1)} \\
&= \varprojlim_I (S_i \otimes_k \mathrm{Hom}_{\mathrm{PC}(C_p)}(C_p/I, S_i)^*) \\
&= S_i
\end{aligned}$$

where as usual I runs through the open twosided ideals in C_p. □

We have the following result.

THEOREM 5.5.9. *The functor* $\hat{(-)}_p$ *defines an equivalence between* $\tilde{C}_{p,f}$ *and* $\mathrm{PCFin}(C_p - C_p)$.

PROOF. First note that thanks to lemma 5.5.8 and Proposition 5.4.4 the image of $\tilde{C}_{p,f}$ under $(-)_p^\wedge$ is contained in $\mathrm{PCFin}(C_p - C_p)$. To show that this is actually an equivalence we will construct an inverse.

Assume $U \in \mathrm{PCFin}(C_p - C_p)$ and $\mathcal{F} \in \mathrm{mod}(X)$. Define

$$T(\mathcal{F}) = (\hat{\mathcal{F}}_p \otimes_{C_p} U)_p^\sim \tag{5.26}$$

Since $\hat{\mathcal{F}}_p$ is finitely generated (Theorem 5.3.1) we find that T defines a right exact additive functor $\mathrm{mod}(X) \to \mathcal{C}_{f,p} \subset \mathrm{mod}(X)$. Hence this functor extends to a functor $\tilde{T} : \mathrm{Mod}(X) \to \mathrm{Mod}(X)$ commuting with colimits. We denote by $\tilde{U}_p$ the right adjoint to this functor. If we view $\tilde{U}_p$ as an object in $\mathrm{Bimod}(o_X - o_X)$ then notationally we have

$$\mathcal{F} \otimes_{o_X} \tilde{U}_p = T(\mathcal{F}) = (\hat{\mathcal{F}}_p \otimes_{C_p} U)_p^\sim \tag{5.27}$$

We now show that $\tilde{(-)}_p$ is a left inverse to $\hat{(-)}_p$. Let $\mathcal{S} \in \mathcal{C}_{f,p}$. To show that $(\hat{\mathcal{S}}_p)_p^\sim = \mathcal{S}$ it suffices to construct a natural isomorphism

$$\mathcal{F} \otimes_{o_X} (\hat{\mathcal{S}}_p)_p^\sim = \mathcal{F} \otimes_{o_X} \mathcal{S}$$

where $\mathcal{F} \in \mathrm{mod}(X)$. With the help of (5.27) we compute

$$\begin{aligned} \mathcal{F} \otimes_{o_X} (\hat{\mathcal{S}}_p)_p^\sim &= (\hat{\mathcal{F}}_p \otimes_{C_p} \hat{\mathcal{S}}_p)_p^\sim \\ &= ((\mathcal{F} \otimes_{o_X} \mathcal{S})_p^\wedge)_p^\sim \qquad \text{(Proposition 5.4.5)} \\ &= \mathcal{F} \otimes_{o_X} \mathcal{S} \end{aligned}$$

Thus the composition

$$\tilde{\mathcal{C}}_{f,p} \xrightarrow{\hat{(-)}_p} \mathrm{PCFin}(C_p - C_p) \xrightarrow{\tilde{(-)}_p} \mathrm{Bimod}(o_X - o_X) \tag{5.28}$$

is the identity on $\tilde{\mathcal{C}}_{f,p}$. We now claim that the essential image of $\tilde{(-)}_p$ is contained in $\tilde{\mathcal{C}}_{f,p}$. Since $\tilde{\mathcal{C}}_{f,p}$ is closed under extensions in $\mathrm{Bimod}(o_X - o_X)$ and since by (5.27) $\tilde{(-)}_p$ is at least right exact (it is of course exact) it suffices to show that $\tilde{S}_{i,p} \in \tilde{\mathcal{C}}_{f,p}$. This follows from the fact that $(o_{\tau^i p})_p^\wedge = S_i$ by lemma 5.5.8, whence $\tilde{S}_{i,p} = o_{\tau^i p}$.

Now we show that $\tilde{(-)}_p$ is also a right inverse to $\hat{(-)}_p$. Let $S \in \mathrm{PCFin}(C_p - C_p)$. We show that $(\tilde{S}_p)_p^\wedge = S$. By definition

$$\begin{aligned} (\tilde{S}_p)_p^\wedge &= \varprojlim_I ((C_p/I)_p^\sim \otimes_{o_X} \tilde{S}_p)_p^\wedge \\ &= \varprojlim_I ((((C_p/I)_p^\sim)_p^\wedge \otimes_{C_p} S)_p^\sim)_p^\wedge \qquad \text{(eq. (5.27))} \\ &= \varprojlim (C_p/I) \otimes_{C_p} S \\ &= S \quad \square \end{aligned}$$

To close this section we discuss $\mathcal{T}or$ between objects of $\tilde{\mathcal{C}}_{f,p}$.

THEOREM 5.5.10. *Let $\mathcal{F} \in \operatorname{mod}(X)$, $\mathcal{T} \in \tilde{\mathcal{C}}_{f,p}$ and let $\mathcal{S}$ be a coherent $\mathcal{C}_p$-preserving o_X-bimodule such that $\hat{\mathcal{S}}_p$ is finitely presented on the right. Then*

$$\mathcal{T}or_i^{o_X}(\mathcal{F},\mathcal{T}) \in \mathcal{C}_{f,p} \tag{5.29}$$

$$\mathcal{T}or_i^{o_X}(\mathcal{S},\mathcal{T}) \in \tilde{\mathcal{C}}_{f,p} \tag{5.30}$$

$$\operatorname{Tor}_i^{\mathrm{PC}(C_p)}(\hat{\mathcal{F}}_p,\hat{\mathcal{T}}_p) \in \operatorname{PCFin}(C_p) \tag{5.31}$$

$$\operatorname{Tor}_i^{\mathrm{PC}(C_p)}(\hat{\mathcal{S}}_p,\hat{\mathcal{T}}_p) \in \operatorname{PCFin}(C_p - C_p) \tag{5.32}$$

Furthermore we have

$$\mathcal{T}or_i^{o_X}(\mathcal{F},\mathcal{T})_p^{\wedge} = \operatorname{Tor}_i^{\mathrm{PC}(C_p)}(\hat{\mathcal{F}}_p,\hat{\mathcal{T}}_p) \tag{5.33}$$

$$\mathcal{T}or_i^{o_X}(\mathcal{S},\mathcal{T})_p^{\wedge} = \operatorname{Tor}_i^{\mathrm{PC}(C_p)}(\hat{\mathcal{S}}_p,\hat{\mathcal{T}}_p) \tag{5.34}$$

PROOF. By Proposition 5.1.6 and Theorem 5.3.1 $\hat{\mathcal{F}}_p$ has a resolution by free modules of finite rank. This implies (5.31). From the hypotheses we also find that $\operatorname{Tor}_i^{\mathrm{PC}(C_p)}(\hat{\mathcal{S}}_p,\hat{\mathcal{T}}_p)$ lies in $\operatorname{PCFin}(C_p)$. Since it also lies in $\mathrm{PC}(C_p - C_p)$, it must necessarily lie in $\operatorname{PCFin}(C_p - C_p)$. This proves (5.32).

To prove the other statements of the theorem it suffices to prove

$$\mathcal{T}or_i^{o_X}(\mathcal{F},\mathcal{T}) = \operatorname{Tor}_i^{\mathrm{PC}(C_p)}(\hat{\mathcal{F}}_p,\hat{\mathcal{T}}_p)_p^{\sim} \tag{5.35}$$

$$\mathcal{T}or_i^{o_X}(\mathcal{S},\mathcal{T}) = \operatorname{Tor}_i^{\mathrm{PC}(C_p)}(\hat{\mathcal{S}}_p,\hat{\mathcal{T}}_p)_p^{\sim} \tag{5.36}$$

Recall that to prove these equations, we have to show that they represent the same functor on injectives.

We'll only prove (5.36). (5.35) is similar.

Let E be an injective object in $\operatorname{Mod}(X)$. Let E_1 be the $\mathcal{C}_p$ part of E. We have

$$\begin{aligned}
&\mathcal{H}om_{\operatorname{Mod}(X)}(\operatorname{Tor}_i^{\mathrm{PC}(C_p)}(\hat{\mathcal{S}}_p,\hat{\mathcal{T}}_p)_p^{\sim}, E)\\
&\quad = \mathcal{H}om_{\operatorname{Mod}(X)}(\operatorname{Tor}_i^{\mathrm{PC}(C_p)}(\hat{\mathcal{S}}_p,\hat{\mathcal{T}}_p)_p^{\sim}, E_1)\\
&\quad = \operatorname{Hom}_{\operatorname{Top}(C_p)}(\operatorname{Tor}_i^{\mathrm{PC}(C_p)}(\hat{\mathcal{S}}_p,\hat{\mathcal{T}}_p), \hat{E}_{1,p})_p^{\sim} \qquad \text{(Prop. 5.4.3)}\\
&\quad = \operatorname{Ext}^i_{\operatorname{Top}(C_p)}(\hat{\mathcal{S}}^p, \operatorname{Hom}_{\operatorname{Dis}(C_p)}(\hat{\mathcal{T}}_p, \hat{E}_{1,p}))_p^{\sim} \qquad \text{(lemma 4.20)}\\
&\quad = \mathcal{E}xt^i_{\operatorname{Mod}(X)}(\mathcal{S}, \mathcal{H}om_{\operatorname{Mod}(X)}(\mathcal{T}, E_1)) \qquad \text{(Prop. 5.4.3)}\\
&\quad = \mathcal{E}xt^i_{\operatorname{Mod}(X)}(\mathcal{S}, \mathcal{H}om_{\operatorname{Mod}(X)}(\mathcal{T}, E))\\
&\quad = \mathcal{H}om_{\operatorname{Mod}(X)}(\mathcal{T}or_i(\mathcal{S},\mathcal{T}), E) \qquad \text{(eq. (3.1))}
\end{aligned}$$

Done! □

5.6. Completion of algebras

In the sequel we will need the following concept.

DEFINITION 5.6.1. A topological $\mathbb{Z}$-graded ring A is said to be graded cofinite if A_0 is cofinite and if the homogeneous parts of A are bipseudo-compact A_0-modules.

If A is as above then we denote by $\operatorname{GrTop}(A)$ the category of graded right topological A-modules. With $\operatorname{GrDis}(A)$, resp. $\operatorname{GrPC}(A)$ we denote the full subcategories consisting of objects $\oplus_m M_m$ such that M_m is in $\operatorname{Dis}(A_0)$, resp. $\operatorname{PC}(A_0)$.

We say that A is (right) locally noetherian if for every indecomposable idempotent in A_0 we have that $e_i A$ is noetherian in $\mathrm{GrPC}(A)$. Left locally noetherian is defined similarly.

If A is locally noetherian then it easily follows that $\mathrm{GrDis}(A)$ is locally noetherian. By analogy with the definition of Proj of an ordinary graded ring we define $\mathrm{Tors}(A)$ as the full subcategory of $\mathrm{GrDis}(A)$ consisting of objects which are direct limits of right bounded ones. We then put $\mathrm{QGrDis}(A) = \mathrm{GrDis}(A)/\mathrm{Tors}(A)$.

Now let X, Y, p etc ... be as in the previous sections. Let $\mathcal{A} = \oplus_n \mathcal{A}_n$ be in $\mathrm{Gralg}(X)$ and assume that the $\mathcal{A}_n$ are coherent and $\mathcal{C}_p$ preserving. Then it makes sense to define $\hat{\mathcal{A}}_p$ as $\oplus_n \hat{\mathcal{A}}_p$. The compatibility of $\hat{(-)}_p$ with tensor product implies that $\hat{\mathcal{A}}_p$ is a cofinite algebra.

As in §3.10 we define $\mathcal{C}_p(\mathcal{A})$ as the graded $\mathcal{A}$-modules $\mathcal{M}$ such $\mathcal{M}_n \in \mathcal{C}_p$ for all n. Similarly $Q\mathcal{C}_p(\mathcal{A})$ is the image of $\mathcal{C}_p(\mathcal{A})$ in $\mathrm{Gr}(\mathcal{A})$.

It is easy to prove the following.

PROPOSITION 5.6.2. *The functor $\hat{(-)}_p$ defines an equivalence between $\mathcal{C}_p(\mathcal{A})$ and $\mathrm{GrDis}(\hat{\mathcal{A}}_p)$ and between $Q\mathcal{C}_p(\mathcal{A})$ and $\mathrm{QGrDis}(\hat{\mathcal{A}}_p)$.*

PROOF. This follows easily from compatibility of completion with tensor product (Proposition 5.4.5). □

5.7. Multiplicities in the case that τ has infinite order

This section is somewhat more special than the previous ones. Our aim is to use the completion functor to attach to certain $\mathcal{F} \in \mathrm{mod}(X)$ an invariant $T_p(\mathcal{F})$ which, in some sense, is an analogue for the multiplicities of $\mathcal{F}$ in p and in the points infinitely near to p. A straightforward treatment seems only to be possible in the case that the τ-orbit of p is infinite, *so we assume this throughout this section.* Note that this implies in particular that p is smooth on Y.

To simplify the notations we will write $\mathcal{F}_Y$ for $i^*\mathcal{F} = \mathcal{F}/\mathcal{F}(-Y)$. We will say that $\mathcal{F} \in \mathrm{mod}(X)$ is Y-transversal if $\mathcal{F}_Y$ has finite length. It then follows automatically that $\ker(\mathcal{F}(-Y) \to \mathcal{F})$ also has finite length (exercise).

The category of Y-transversal objects is denoted by $\mathrm{trans}_Y(X)$. If $\mathcal{F}$ is an object in $\mathrm{trans}_Y(X)$ then, in a slight generalization of [**3, 1**], we define $\mathrm{Div}(\mathcal{F})$ as $[\mathcal{F}_Y] - [\ker(\mathcal{F}(-Y) \to \mathcal{F})]$ where $[-]$ denotes the effective divisor associated to a finite length module.

It is easy to see that Div is additive on short exact sequences. Furthermore if one has an epimorphism $\mathcal{F} \to \mathcal{O}_q$, with kernel $\mathcal{G}$ then the following formula holds [**1**]

$$\mathrm{Div}(\mathcal{G}) = \mathrm{Div}(\mathcal{F}) - (q) + (\tau^{-1}q) \tag{5.37}$$

From this we see that (in contrast to the commutative case) the invariant $\mathrm{Div}(\mathcal{F})$ depends on $\mathcal{F}$ itself and not only on the image of $\mathcal{F}$ in $\mathrm{mod}(X)/\mathcal{C}_f$. We now define a better behaving invariant.

Assume that $\mathcal{F} \in \mathrm{trans}_Y(X)$. We look at the exact sequence

$$0 \to \mathcal{G} \to \mathcal{F}(-Y) \to \mathcal{F} \to \mathcal{H} \to 0 \tag{5.38}$$

By definition $\mathcal{G}$, $\mathcal{H}$ are finite length modules on Y.

The completion of (5.38) looks like

$$0 \to \hat{\mathcal{G}}_p \to \hat{\mathcal{F}}_{p,\phi} \xrightarrow{\times N} \hat{\mathcal{F}}_p \to \hat{\mathcal{H}}_p \to 0 \tag{5.39}$$

Now $\hat{\mathcal{G}}_p$, $\hat{\mathcal{H}}_p$ are finite length modules over $C_p/(N)$. It follows from Nakayama's lemma(lemma 4.8) that $\hat{\mathcal{F}}_p$ is a quotient of a finite direct sum of P_i's.

Using (4.11) we can decompose (5.39) as a product of exact sequences of the form

$$0 \longrightarrow \hat{\mathcal{G}}_p e_i \longrightarrow \hat{\mathcal{F}}_p e_{i+1} \xrightarrow{\times N} \hat{\mathcal{F}}_p e_i \longrightarrow \hat{\mathcal{H}}_p e_i \longrightarrow 0 \tag{5.40}$$

Since $\hat{\mathcal{F}}_p$ is a quotient of a finite direct sum of P_i's it follows that $\hat{\mathcal{F}}_p e_i = 0$ will be zero if $i \gg 0$. Therefore it follows from (5.40) that $\hat{\mathcal{F}}_p e_i$ is a finite dimensional R-module. Furthermore since $\hat{\mathcal{G}}_p e_i$, $\hat{\mathcal{H}}_p e_i$ are only non-zero for a finite number of i, it follows that $\hat{\mathcal{F}}_p e_{i+1} \xrightarrow{\times N} \hat{\mathcal{F}}_p e_i$ is an isomorphism for large negative i. We define $T_p(\mathcal{F})$ as $\varprojlim_i \hat{\mathcal{F}}_p e_i$ (with transition maps given by N). Thus T_p is a functor from $\operatorname{trans}_Y(X)$ to finite dimensional R-modules.

PROPOSITION 5.7.1. *T_p has the following properties.*

1. *T_p is exact.*
2. *$T_p(\mathcal{F})$ does only depend on the image of $\mathcal{F}$ in* $\operatorname{mod}(X)/\mathcal{C}_{f,p}$.

PROOF. 1. This follows from the exactness of the completion functor.
2. This follows from the fact that for a finite dimensional C_p-module G one has $Ge_i = 0$ for almost all i. □

Let us now indicate a direct way of computing $T_p(\mathcal{F})$. We say that an object $\mathcal{F}$ is (p-)*normalized* if $\mathcal{F}$ is $\mathcal{C}_{f,p}$-torsion free and $\tau^i p$ is not in the support of $\mathcal{F}_Y = 0$, unless $i = 0$. We have the following result.

PROPOSITION 5.7.2. 1. *Every $\mathcal{F} \in \operatorname{trans}_Y(X)$ is equivalent modulo $\mathcal{C}_{f,p}$ to a unique p-normalized object in* $\operatorname{mod}(X)$.
2. *If $\mathcal{F}$ is p-normalized then $T_p(\mathcal{F}) = \hat{\mathcal{F}}_{Y,p}$.*

PROOF. 1. Let us first consider uniqueness. If we have an inclusion of p-normalized objects $\mathcal{F} \subset \mathcal{F}'$ such that $\mathcal{F}'/\mathcal{F} \in \mathcal{C}_{f,p}$ then by (5.37) $\mathcal{F} = \mathcal{F}'$. Now assume that $\mathcal{F}$ and $\mathcal{F}'$ are p-normalized objects which are equivalent modulo $\mathcal{C}_{f,p}$. This means that there exist a $\mathcal{C}_{f,p}$-torsion free $\mathcal{G}$, together with inclusions

$$\mathcal{F} \rightarrow \mathcal{G} \leftarrow \mathcal{F}'$$

whose cokernel lies in $\mathcal{C}_{f,p}$. We now replace $\mathcal{F}$, $\mathcal{F}'$ by their images in $\mathcal{G}$ and we put $\mathcal{H} = \mathcal{F} + \mathcal{F}'$. Since $\mathcal{H}$ is a quotient of $\mathcal{F} \oplus \mathcal{F}'$ it follows that $\mathcal{H}$ is also p-normalized. But then by the above it follows that $\mathcal{F} = \mathcal{H} = \mathcal{F}'$.

Let us now consider existence. Assume that $\mathcal{F} \in \operatorname{trans}_Y(X)$. Without loss of generality we may assume that $\mathcal{F}$ is $\mathcal{C}_{f,p}$-torsion free. We will then modify $\operatorname{Div}(\mathcal{F})$ step by step until it is p-normalized.

If $\tau^i p \in \operatorname{Div}(\mathcal{F})$ with $i > 0$ then we let $\mathcal{F}'$ be the kernel of the associated epimorphism $\mathcal{F} \rightarrow \mathcal{O}_{\tau^i p}$. By (5.37) it follows that $\mathcal{F}'$ is closer to being p-normalized than the original $\mathcal{F}$.

Assume now $\tau^i p \in \operatorname{Div}(\mathcal{F})$ with $i < 0$. In this case we use the associated monomorphism $\mathcal{O}_{\tau^{i+1}p} \to \mathcal{F}_Y(Y)$ and we let $\mathcal{F}'$ be the pullback of the maps

$$\begin{array}{ccc} \mathcal{F}(Y) & \longrightarrow & \mathcal{F}_Y(Y) \\ & & \uparrow \\ & & \mathcal{O}_{\tau^{i+1}p} \end{array}$$

One now easily checks

$$\operatorname{Div}(\mathcal{F}') = \operatorname{Div}(\mathcal{F}) - (\tau^i p) + (\tau^{i+1} p)$$

so that we have again made progress. Repeating these constructions we eventually find a normalized object, which is equivalent to $\mathcal{F}$.

2. Assume that $\mathcal{F}$ is normalized. Then $\hat{\mathcal{F}}_{Y,p} e_i$ is zero, except if $i = 0$. Whence we find from the exact sequence (5.40) $T_p(\mathcal{F}) = \hat{\mathcal{F}}_{Y,p} e_0 = \hat{\mathcal{F}}_{Y,p}$. □

Let N_p be the functor which associates to $\mathcal{F} \in \operatorname{trans}_Y(X)$ its normalization. Then we have proved the following

COROLLARY 5.7.3. *The functor N_p defines an equivalence of categories between* $\operatorname{trans}_Y(X)/\mathcal{C}_{f,p}$ *and the full subcategory of* $\operatorname{trans}_Y(X)$ *consisting of p-normalized objects.*

If $\mathcal{F}$ is p-normalized then we can reconstruct $\hat{\mathcal{F}}_p$ from $T_p(\mathcal{F})$.

LEMMA 5.7.4. *Assume that $\mathcal{F} \in \operatorname{trans}_Y(X)$ is p-normal. Then $\hat{\mathcal{F}}_p$ is isomorphic to the row vector*

$$(\cdots T_p(\mathcal{F}) \cdots T_p(\mathcal{F})\ 0 \cdots 0 \cdots)$$

where the right most $T_p(\mathcal{F})$ occurs in position 0.

PROOF. Apriori $\hat{\mathcal{F}}_p$ is given by the row vector $(\hat{\mathcal{F}}_p e_i)_i$ (4.11). By the previous discussion $\hat{\mathcal{F}}_p e_i = 0$ for $i \gg 0$ and multiplication by N is an isomorphism on $\hat{\mathcal{F}}_p e_i$ for $i \neq 0$. It now follows from the definition that

$$\hat{\mathcal{F}}_p e_i = \begin{cases} T_p(\mathcal{F}) & \text{if } i \leq 0 \\ 0 & \text{if } i > 0 \end{cases} \qquad \square$$

CHAPTER 6

Blowing up a point on a commutative divisor

6.1. Some ideals

Let X, Y, p be as above and let q be an arbitrary point on Y. The bimodule o_q defined above is by construction a quotient of o_Y and hence also of o_X. We define

$$\begin{aligned} m_{Y,q} &= \ker(o_Y \to o_q) \\ m_q &= \ker(o_X \to o_q) \end{aligned}$$

By Corollary 5.5.6, $m_{Y,q}$, $m_q \in \mathrm{Bimod}(o_X - o_X)$. We also define

$$\begin{aligned} I_Y &= m_{Y,p} \otimes_{o_X} o_X(Y) \\ &= m_{Y,p} \otimes_{o_Y} \mathcal{N}_{Y/X} \\ I &= m_p \otimes_{o_X} o_X(Y) \end{aligned} \tag{6.1}$$

Clearly $I_Y \subset \mathcal{N}_{Y/X}$, $I \subset o_X(Y)$. Below we give some properties of $I^n \overset{\text{def}}{=} \mathrm{im}(I^{\otimes n} \to o_X(nY))$. It will be clear that suitable analogs for I_Y^n hold.

PROPOSITION 6.1.1. 1. $o_X(nY)/I^n \in \tilde{\mathcal{C}}_{f,p}$.
2. *I^n is a $\mathcal{C}_p$-preserving coherent bimodule contained in* $\mathrm{Bimod}(o_X - o_X)$.

PROOF. 1. We use induction on n, the case $n = 1$ being clear. Put $\mathcal{S}_n = o_X(nY)/I^n$. Tensoring the exact sequence

$$0 \to I^n \to o_X(nY) \to \mathcal{S}_n \to 0 \tag{6.2}$$

with I yields an exact sequence

$$I \otimes_{o_X} I^n \to I \otimes_{o_X} o_X(nY) \to I \otimes_{o_X} \mathcal{S}_n \to 0$$

and thus

$$I^{n+1} = \ker(I \otimes_{o_X} o_X(nY) \to I \otimes_{o_X} \mathcal{S}_n) \tag{6.3}$$

Viewing $I \otimes o_X(nY)$ and I^{n+1} as subbimodules of $o_X((n+1)Y)$ we obtain from (6.3) an exact sequence

$$0 \to I \otimes_{o_X} \mathcal{S}_n \to \mathcal{S}_{n+1} \to \mathcal{S}_1(nY) \to 0$$

Hence it suffices to prove that $I \otimes_{o_X} \mathcal{S}_n \in \tilde{\mathcal{C}}_{p,f}$.

Tensoring

$$0 \to I \to o_X(Y) \to o_{\tau p} \to 0$$

with $\mathcal{S}_n$ yields

$$0 \to \mathcal{T}or_1^{o_X}(o_{\tau p}, \mathcal{S}_n) \to I \otimes_{o_X} \mathcal{S}_n \to o_X(Y) \otimes_{o_X} \mathcal{S}_n \to o_{\tau p} \otimes_{o_X} \mathcal{S}_n \to 0$$

By (5.30) it follows that indeed $I \otimes_{o_X} \mathcal{S}_n \in \tilde{\mathcal{C}}_{f,p}$.

2. From the exact sequence (6.2), Proposition 3.1.8 and Corollary 5.5.6 it follows that indeed I^n is coherent and is contained in $\mathrm{Bimod}(o_X - o_X)$.

   Let $\mathcal{T} \in \mathcal{C}_p$. Applying $\mathcal{T} \otimes_{o_X} -$ and $\mathcal{H}om_{o_X}(-, \mathcal{T})$ to the sequence (6.2) yields long exact sequences

$$0 \to \mathcal{T}or_1^{o_X}(\mathcal{T}, \mathcal{S}_n) \to \mathcal{T} \otimes_{o_X} I^n \to \mathcal{T}(nY) \to \mathcal{T} \otimes_{o_X} \mathcal{S}_n \to 0$$
$$0 \to \mathcal{H}om_{o_X}(\mathcal{S}_n, \mathcal{T}) \to \mathcal{T}(-nY) \to \mathcal{H}om_{o_X}(I^n, \mathcal{T}) \to \mathcal{E}xt^1_{o_X}(\mathcal{S}_n, \mathcal{T}) \to 0$$

   Thanks to (5.29) and Proposition 5.4.3 we can conclude that I^n is $\mathcal{C}_p$ preserving.

The following proposition gives a little additional information on I^n which we will need below.

PROPOSITION 6.1.2. *I^n has the following additional properties. Assume that $\mathcal{M} \in \mathrm{mod}(X)$. Then*

$$\mathcal{T}or_i^{o_X}(\mathcal{M}, I^n) \quad \begin{cases} \in \mathrm{mod}(X) & \text{if } i = 0 \\ \in \mathcal{C}_{f,p} & \text{if } i = 1 \\ = 0 & \text{if } i \geq 2 \end{cases}$$

PROOF. The case $i = 0$ is already covered by Proposition 6.1.1. Hence we consider the case $i > 0$. Tensoring the exact sequence (6.2) on the left with $\mathcal{M}$ yields

$$\mathcal{T}or_i^{o_X}(\mathcal{M}, I^n) = \mathcal{T}or_{i+1}^{o_X}(\mathcal{M}, \mathcal{S}_n)$$

Hence by Theorem 5.5.10 $\mathcal{T}or_i^{o_X}(\mathcal{M}, I^n) \in \mathcal{C}_{f,p}$ for $i \geq 1$.

It remains to be shown that $\mathcal{T}or_j^{o_X}(\mathcal{M}, \mathcal{S}_n) = 0$ for $j \geq 3$. Using Lemma 5.5.8 and Theorem 5.5.10, it is sufficient to know that the left projective dimension of S_i is ≤ 2. This is the version for left modules of Proposition 5.1.7. □

The following proposition gives a more explicit description of I^n.

PROPOSITION 6.1.3. *One has the following alternative expression for I^n.*

$$I^n = m_p m_{\tau^{-1}p} \cdots m_{\tau^{n-1}p} \otimes_{o_X} o_X(nY)$$

We have

$$(I^n)_p^\wedge = \hat{I}_p^n \tag{6.4}$$
$$= (m_0 m_{-1} m_{-2} \cdots m_{-n+1})_{\phi^{-n+1}} \tag{6.5}$$

as subobjects of the invertible C_p-bimodule $(C_p)_{\phi^{-n+1}}$. Here ϕ is as in Theorem 5.1.4 and for $i \in \mathbb{Z}$, m_i is the twosided maximal ideal in C_p corresponding to S_i.

PROOF. The first statement follows easily from (5.23) which implies that

$$o_X(Y) \otimes_{o_X} m_q = m_{\tau^{-1}q} \otimes_{o_X} o_X(Y)$$

(as subobjects of $o_X(Y)$).

Now we prove the second statement. Completing the exact sequence

$$0 \to m_p \to o_X \to o_p \to 0$$

and using lemma 5.5.8 gives $\hat{m}_p = m_0$. Furthermore applying definition (5.16) together with Theorem 5.1.4.4 yields $o_X(Y)_p^\wedge = (C_p)_{\phi^{-1}}$. Using the compatibility of tensor product with $(-)_p^\wedge$ we then deduce from (6.1)

$$\hat{I}_p = (m_0)_{\phi^{-1}} \tag{6.6}$$

(6.4) is easily proved by induction. By (6.3) we have an exact sequence

$$0 \to I^{n+1} \to I \otimes_{o_X} o_X(nY) \to I \otimes_{o_X} \mathcal{S}_n \to 0$$

which yields an exact sequence (using exactness of $\hat{(-)}$ and compatibility with tensor product, see Propositions 5.4.4 and 5.4.5)

$$0 \to (I^{n+1})_p^\wedge \to (\hat{I}_p)_{\phi^{-n}} \to \hat{I}_p \otimes_{o_X} \hat{\mathcal{S}}_{n,p} \to 0 \tag{6.7}$$

On the other hand completing the exact sequence

$$0 \to I^n \to o_X(nY) \to \mathcal{S}_n \to 0$$

and tensoring with $\hat{I}_p$ yields by induction an exact sequence

$$0 \to \hat{I}_p \cdot \hat{I}_p^n \to (\hat{I}_p)_{\phi^{-n}} \to \hat{I}_p \otimes_{o_X} \hat{\mathcal{S}}_{n,p} \to 0 \tag{6.8}$$

Comparing (6.7) and (6.8) yields what we want.

(6.5) now follows from (6.4),(6.6) and the easily verified fact that $\phi^{-1}(m_i) = m_{i-1}$. □

Before we continue we make a few remarks on the case that p is a fixed point for τ. In that case it follows from (5.23) that

$$m_p \otimes_{o_X} o_X(Y) = o_X(Y) \otimes_{o_X} m_p \tag{6.9}$$

as subobjects of $o_X(Y)$.

Let μ be the multiplicity of p in Y. If p is a fixed point then it is possible that $\mu \geq 1$ by Theorem 5.1.4. We have the following description of μ which is completely analogous to the classical case.

LEMMA 6.1.4. *Assume that p is a fixed point for τ. Then μ is the largest integer n such that $o_X(-Y) \subset m_p^n$ as subbimodules of o_X.*

PROOF. We have to find the largest integer such that

$$o_X/m_p^n \to o_X/(o_X(-Y) + m_p^n)$$

is an isomorphism. Using the properties of $\hat{(-)}$ and in particular Theorem 5.5.9 we see that this is equivalent to

$$R/m^n \to R/(RU + m^n)$$

being an isomorphism, where m is the maximal ideal of R. This in turn is the same as saying that $U \in m^n$. Now lemma 5.1.8 yields what we want. □

PROPOSITION 6.1.5. *Let μ be the multiplicity of Y in p and let $n \in \mathbb{Z}$. Then the natural exact sequence*

$$0 \to o_X((n-1)Y) \to o_X(nY) \to o_Y(nY) \to 0$$

restricts to an exact sequence

$$0 \to I^{n-\mu}((\mu-1)Y) \to I^n \to I_Y^n \to 0 \tag{6.10}$$

Here I^n for $n < 0$ is to be interpreted as $o_X(nY)$.

PROOF. Note that by (6.9) there is no ambiguity in the notation $I^{n-\mu}((\mu-1)Y)$.

First we verify that the lower sequence is well defined. We have

$$\mathcal{O}(-Y) \subset m_p m_{\tau^{-1}p} \cdots m_{\tau^{\mu-1}p} \tag{6.11}$$

If $\mu = 1$ then this is by definition. If $\mu > 1$ then p is a fixed point and we can use lemma 6.1.4. Tensoring (6.11) with $o_Y(\mu Y)$ yields an inclusion

$$\mathcal{O}((\mu-1)Y) \subset I^{\mu}$$

Assume $n \geq \mu$. Multiplying with $I^{n-\mu}$ then yields an inclusion

$$I^{n-\mu}((\mu-1)Y) \subset I^n \tag{6.12}$$

In a similar way one verifies directly from (6.11) that (6.12) also holds if $n < \mu$. Hence (6.10) is indeed well defined.

To prove the proposition we look at the following commutative diagram.

$$\begin{array}{ccccccccc}
 & & 0 & & 0 & & 0 & & \\
 & & \downarrow & & \downarrow & & \downarrow & & \\
0 & \longrightarrow & I^{n-\mu}((\mu-1)Y) & \xrightarrow{\alpha} & I^n & \xrightarrow{\beta} & I^n_Y & \longrightarrow & 0 \\
 & & \downarrow & & \downarrow & & \downarrow & & \\
0 & \longrightarrow & o_X((n-1)Y) & \longrightarrow & o_X(nY) & \longrightarrow & o_Y(nY) & \longrightarrow & 0 \\
 & & \downarrow & & \downarrow & & \downarrow & & \\
 & & \mathcal{S}_{n-\mu}((\mu-1)Y) & \xrightarrow{\gamma} & \mathcal{S}_n & \xrightarrow{\delta} & \mathcal{S}_{Y,n} & \longrightarrow & 0 \\
 & & \downarrow & & \downarrow & & \downarrow & & \\
 & & 0 & & 0 & & 0 & &
\end{array}$$

where as usual $\mathcal{S}_n = o_X(nY)/I^n$, $\mathcal{S}_{Y,n} = o_Y(nY)/I^n_{Y,p}$. If $n < 0$ then $\mathcal{S}_n$, $\mathcal{S}_{n,Y}$ are defined as zero.

In this diagram the middle row is exact, α is monic and β is an epimorphism. Elementary diagram chasing shows that the lower row is exact.

We now have to show that $\ker\beta = \operatorname{im}\alpha$. A diagram chase shows that this is equivalent to γ being monic. This is clear if $n < \mu$ so we may assume that $n \geq \mu$. Injectivity of γ is then in turn equivalent to

$$\operatorname{length}\mathcal{S}_n = \operatorname{length}\mathcal{S}_{n-\mu} + \operatorname{length}\mathcal{S}_{Y,n} \tag{6.13}$$

By Theorem 5.5.9 this can be checked on C_p. By Corollary 5.2.4 we have

$$\operatorname{length}\mathcal{S}_n = \frac{n(n+1)}{2}$$

If $|O_\tau(p)| > 1$ then by Theorem 5.1.4 p is smooth on Y and hence

$$\operatorname{length}\mathcal{S}_{Y,n} = n$$

One sees that in this case (6.13) is satisfied.

Assume now that $\tau p = p$. By Theorem 5.5.9 we may check the injectivity of γ after completing. We have $C_p = R$ as in Theorem 5.1.4. Let m be the maximal ideal of R. We then have $\hat{\mathcal{S}}_n = R/m^n$ and the map $\hat{\gamma}_p$ is given by right multiplication by U. From lemma 5.1.8 it follows that $U \in m^\mu - m^{\mu+1}$. From this and the fact that

$\operatorname{gr} R$ is a domain (for the m-adic filtration) we deduce that right multiplication by U defines an injective map $R/m^{n-\mu} \to R/m^n$. □

6.2. Some Rees algebras

We will use I_Y, I to define the following Rees algebras.

$$\mathcal{D}_Y = o_Y \oplus I_Y \oplus I_Y^2 \oplus \cdots$$
$$\mathcal{D} = o_X \oplus I \oplus I^2 \oplus \cdots$$

As a corollary to Proposition 6.1.1 we immediately deduce

COROLLARY 6.2.1. *$\mathcal{D}_Y, \mathcal{D} \in \operatorname{Alg}(X)$.*

Now we deduce some other good properties of $\mathcal{D}$. We denote shifting in $\operatorname{Gr}(\mathcal{D})$ and $\operatorname{Bigr}(\mathcal{D})$ by $(-)$.

As usual the case where p is fixed point for τ is somewhat peculiar. In that case it follows easily from (6.9) that

$$\mathcal{D}(nY) \overset{\text{def}}{=} \mathcal{D} \otimes_{o_X} o_X(nY) = o_X(nY) \otimes_{o_X} \mathcal{D} \tag{6.14}$$

is a twosided invertible graded bimodule over $\mathcal{D}$. This is clearly false if $\tau p \neq p$.

It is sometimes convenient to define a modified Rees algebra $\tilde{\mathcal{D}}$ by

$$\tilde{\mathcal{D}}_n = \begin{cases} o_X(nY) & \text{if } n < 0 \\ I^n & \text{if } n \geq 0 \end{cases} \tag{6.15}$$

$\tilde{\mathcal{D}}_Y$ is defined similarly but with o_Y, I_Y replacing o_X, I.

THEOREM 6.2.2. *Let $\mathcal{D}$, $\mathcal{D}_Y$ be as above. Let μ be the multiplicity of p on Y. Then*

1. *There is an exact sequence of graded $\tilde{\mathcal{D}}$-bimodules.*

$$0 \to \tilde{\mathcal{D}}(-\mu)((\mu-1)Y) \to \tilde{\mathcal{D}} \to \tilde{\mathcal{D}}_Y \to 0 \tag{6.16}$$

2. *$\mathcal{D}$ is noetherian.*
3. *$\mathcal{D}$ satisfies χ.*
4. *$\operatorname{cd} \tau_{\mathcal{D}} \leq 2$.*

PROOF. The fact that (6.16) exists follows directly from Proposition 6.1.5.

Our aim is now to deduce 2., 3. and 4. from the corresponding statements for $\mathcal{D}_Y$. So we first have to handle this case. If $\mu = 1$ then I_Y is an invertible bimodule. Then we can apply Proposition 3.8.13 to the exact sequence

$$0 \to I_Y \otimes_{o_X} \mathcal{D}_Y(-1) \to \mathcal{D}_Y \to o_Y \to 0$$

and we are done. Hence by Theorem 5.1.4 we only have to treat the case that p is a fixed point. But this means that $m_{Y,p}$ is τ-invariant. Let $\mathcal{E}_Y$ be the ordinary rees algebra associated to $m_{Y,p}$. $\mathcal{E}_Y$ may be viewed as a sheaf of algebras in the ordinary sense and the reader may verify that it satisfies properties 2., 3. and $\operatorname{cd} \tau_{\mathcal{E}_Y} \leq 1$. Furthermore there is a category equivalence given by

$$\operatorname{Gr}(\mathcal{D}_Y) \to \operatorname{Gr}(\mathcal{E}_Y) : \oplus_n \mathcal{M}_n \mapsto \oplus_n \mathcal{M}_n(-nY) \tag{6.17}$$

From this it is routine to pull the good properties of $\mathcal{E}_Y$ back to $\mathcal{D}_Y$.

Define now $\mathcal{D}'_Y$ by

$$(\mathcal{D}'_Y)_n = \begin{cases} \mathcal{D}_Y & \text{if } n \geq \mu \\ \mathcal{D} & \text{otherwise} \end{cases} \tag{6.18}$$

Then (6.16) gives an exact sequence

$$0 \to \mathcal{D}(-\mu)((\mu - 1)Y) \to \mathcal{D} \to \mathcal{D}'_Y \to 0 \tag{6.19}$$

It is easy to see that the fact that $\mathcal{D}_Y$ is noetherian, together with the fact that I^n is coherent for $n = 0, \dots, \mu - 1$, implies that $\mathcal{D}'_Y$ is noetherian. Furthermore $\mathcal{D}_Y$ and $\mathcal{D}'_Y$ have the same tails. So from lemma 3.8.16 we obtain that $\mathcal{D}'_Y$ satisfies χ and furthermore $\operatorname{cd} \tau_{\mathcal{D}'_Y} = \operatorname{cd} \tau_{\mathcal{D}_Y} = 1$.

Now applying Proposition 3.8.13 to (6.19) implies 2., 3. and 4. □

REMARK 6.2.3. The inequality in 6.2.2.4 is of course an equality. However we won't need this.

6.3. Definition of blowing up

We define $\tilde{Y} = \operatorname{Proj} \mathcal{D}_Y$, $\tilde{X} = \operatorname{Proj} \mathcal{D}$ and we call $\tilde{X}$ the blowing up of X in p. We denote by π resp. π_Y the quotient maps $\operatorname{Gr}(\mathcal{D}) \to \operatorname{Qgr}(\mathcal{D})$ resp. $\operatorname{Gr}(\mathcal{D}_Y) \to \operatorname{Qgr}(\mathcal{D}_Y)$.

It follows from (6.16) that we have a commutative diagram of quasi-schemes

$$\begin{array}{ccc} \tilde{Y} & \xrightarrow{\beta} & Y \\ {\scriptstyle j}\downarrow & & \downarrow{\scriptstyle i} \\ \tilde{X} & \xrightarrow{\alpha} & X \end{array} \tag{6.20}$$

where i is as before, j comes from the quotient map $\mathcal{D} \to \mathcal{D}_Y$ (through Proposition (3.8.11)) and $\alpha^*\mathcal{M} = \pi(\mathcal{M} \otimes_{o_X} \mathcal{D})$, with the analogous definition for β^*.

The derived functors $R^i\alpha_*$ are defined as usual. $L_i\alpha_*$ is defined as in Section §3.9. From lemma 3.9.2 together with Proposition 6.1.2 it follows that $L_i\alpha^*$ actually defines a functor $\operatorname{Mod}(X) \to \operatorname{Mod}(\tilde{X})$.

The following theorem summarizes some of the main properties of blowing up.

THEOREM 6.3.1.
1. *The pair $(\tilde{Y}, \beta)$ is isomorphic in* Qsch/Y *to the ordinary commutative blowing up at p of Y. In particular β is an isomorphism if Y is smooth in p.*
2. *$\tilde{X}$ is a noetherian quasi-scheme.*
3. *The ideal in $o_{\tilde{X}}$ defined by $\tilde{Y}$ is invertible.*
4. *$i_* \circ R^i\beta_* = R^i\alpha_* \circ j_*$.*
5. *We have $\operatorname{cd} \alpha^* \leq 1$, $\operatorname{cd} \alpha_* \leq 1$*
6. *α is proper*
7. *$o_{\tilde{X}}(\tilde{Y})$ is relatively ample for α.*
8. *j makes $\tilde{Y}$ into a divisor in $\tilde{X}$ in the sense of enriched quasi-schemes (cfr §3.6).*

PROOF. These properties are either straightforward translations of the definitions or they follow from properties of $\mathcal{D}$ which we have already proved.

1. As usual we separate two cases. If $\tau p = p$ then we let $\mathcal{E}_Y$ be as in the proof of Theorem 6.2.2. The category equivalence between $\mathrm{Gr}(\mathcal{D}_Y)$ and $\mathrm{Gr}(\mathcal{E}_Y)$ yields in that case what we want.

   If p is smooth on Y (in particular if $\tau p \neq p$) then $m_{Y,p}$ is invertible. Thus the same holds for I_Y and we have

$$\mathcal{D}_Y = o_Y \oplus I_Y \oplus I_Y^{\otimes 2} \cdots$$

   It now follows from Propositions 3.11.4 (with $\mathcal{S} = 0$) that $\mathrm{QGr}(\mathcal{D}_Y) \cong \mathrm{Mod}(Y)$.

   For use below we state the following formula for $\mathcal{N} \in \mathrm{Gr}(\mathcal{D}_Y)$.

$$\beta_*(\pi_Y \mathcal{N}) = \varinjlim \mathcal{N}_n \otimes_{o_Y} I_Y^{\otimes -n} \tag{6.21}$$

   We leave the easy proof to the reader.
2. This follows from Theorem 6.2.2.2.
3. Using Theorem 6.2.2.1. one easily checks that

$$\pi\mathcal{M} \otimes_{o_{\tilde{X}}} o_{\tilde{X}}(-\tilde{Y}) = \pi(\mathcal{M} \otimes_{\mathcal{D}} \mathcal{D}(-\mu)((\mu-1)Y)) \tag{6.22}$$

   What we want to prove now follows from the fact that $\mathcal{D}(-\mu)((\mu-1)Y)$ is an invertible graded bimodule over $\mathcal{D}$ (this is trivial if $\mu = 1$ and if $\mu \neq 1$ it follows from (6.14)).
4. Define $\gamma = i \circ \beta$. Since i_* is exact we have $R^i\gamma_* = i_* \circ R^i\beta_*$. Hence we have to show that

$$R^i\gamma_* = R^i\alpha_* \circ j_* \tag{6.23}$$

   We have $\tilde{Y} = \mathrm{Proj}\,\mathcal{D}'_Y$ where $\mathcal{D}'_Y$ is as in (6.18). (6.19) shows that $\mathcal{D}'_{Y,p}$ satisfies the conditions for Proposition 3.8.12. Hence we can employ that proposition to obtain what we want.
5. This follows from Theorem 6.2.2.4, equation (3.29) and Proposition 6.1.2.
6. This follows from Theorem 6.2.2.3 together with part of Proposition 3.8.7.
7. Since $\mathcal{D}$ satisfies χ it follows from Proposition 3.8.7 that the canonical shift functor $\mathcal{M} \mapsto \mathcal{M}(1)$ is relatively ample. Denote the corresponding bimodule by $o_{\tilde{X}}(1)$. If p is smooth on Y then it follows from (6.22) that $o_{\tilde{X}}(\tilde{Y}) = o_{\tilde{X}}(1)$. So in that case we are done. If p is singular on Y then by (6.14) $- \otimes_{o_X} o_X(Y)$ defines an autoequivalence of $\mathrm{Mod}(\tilde{X})$. Then (6.22) yields

$$o_{\tilde{X}}(\tilde{Y}) = o_{\tilde{X}}(\mu) \otimes_{o_X} o_X((1-\mu)Y) \tag{6.24}$$

   This yields what we want since it is clear that $- \otimes_{o_X} o_X(Y)$ commutes with $R^i\alpha_*$ and $L_i\alpha^*$. Note in passing that (6.24) also makes sense in the case that $\tau p \neq p$ since then $\mu = 1$.
8. Given 3. we only have to show that the map $\mathcal{O}_{\tilde{X}}(-\tilde{Y}) \to \mathcal{O}_{\tilde{X}}$ is monic. Now $\mathcal{O}_{\tilde{X}}(-\tilde{Y})$ is by definition $\alpha^*\mathcal{O}_X \otimes_{o_{\tilde{X}}} o_{\tilde{X}}(-\tilde{Y})$ which according to (6.24) is equal to $\pi(\mathcal{O}_X \otimes_{o_X} \mathcal{D}(-\mu)((\mu-1)Y))$. Hence by (6.16) it suffice to show that $\mathcal{T}or^i_{o_X}(\mathcal{O}_X, I_Y^n((\mu-1)Y)) = 0$ for $i > 0$, $n \geq 0$. As in the commutative case one checks that $\mathcal{T}or^i_{o_X}(\mathcal{O}_X, I_Y^n((\mu-1)Y)) = \mathcal{T}or^i_{o_Y}(\mathcal{O}_Y, I_Y^n((\mu-1)Y))$. One now easily shows that $\mathcal{T}or^i_{o_Y}(\mathcal{O}_Y, I_Y^n((\mu-1)Y)) = 0$, for example using the fact that $I_Y^n \subset o_Y(nY)$ with the quotient being in $\mathcal{C}_{f,p}$, together with the definition of $\mathcal{T}or(-,-)$ (cfr. (3.1)). □

6.4. The normal bundle

The main result of this section (equation (6.26)) will be used in §6.5.

If $t : U \to V$ is a morphism of schemes, $\mathcal{N}$ is a line bundle on Y and $\mathcal{F} \subset \mathcal{N}$ is a quasicoherent subsheaf then $t^{-1}(\mathcal{F})$ is defined as the image of $t^*(\mathcal{F})$ in $t^*(\mathcal{N})$.

Now let us revert to the notations in use in the previous sections. Recall that according to (5.2) we have $\mathcal{N}_{Y/X} = \mathcal{N}_\tau$ where $\mathcal{N}$ is an invertible sheaf on Y and τ is an automorphism of Y.

Define $\tau' : \tilde{Y} \to \tilde{Y}$ as follows. If p is a fixed point for τ then τ extends in a natural way to the blowup of Y at p. We denote this extended morphism by τ'. If p is not a fixed point then the morphism $\beta : \tilde{Y} \to Y$ is an isomorphism and we put $\tau' = \beta^{-1}\tau\beta$. So in all cases we have

$$\beta\tau' = \tau\beta \tag{6.25}$$

Our aim in this section is to prove the following formula

$$\mathcal{N}_{\tilde{Y}/\tilde{X}} = \beta^{-1}(m_{Y,p}\mathcal{N})_{\tau'} \tag{6.26}$$

In view of the commutative case this formula seems quite logical. However there are some pitfalls. The main problem is that a priori $\tilde{Y}$ is only a quasi-scheme. Thus, although we can use the ordinary definition of $\beta^{-1}(m_{Y,p}\mathcal{N})$, the fact that we can consider the result as a bimodule, depends on the "accidental" event that $\tilde{Y}$ is commutative. So to make sense of (6.26) we have to bring in explicitly the identification of $\tilde{Y}$ with a commutative scheme (which was given in the proof of Theorem 6.3.1.1). Below we give the necessary computations. The reader is advised to skim through the rest of this section.

It is most convenient to separate two cases, depending on whether p is a fixed point or not.

p is not a fixed point for τ. In this case $\mu = 1$ and $m_{Y,p}$ is invertible. Thus $\beta^{-1}(m_{Y,p}\mathcal{N}) = \beta^*(m_{Y,p}\mathcal{N})$. We have to show that for $\mathcal{M} \in \mathrm{Mod}(\tilde{Y})$ we have

$$\mathcal{M} \otimes_{o_{\tilde{Y}}} o_{\tilde{Y}}(\tilde{Y}) = \mathcal{M} \otimes_{o_{\tilde{Y}}} (\beta^*(m_{Y,p}\mathcal{N}))_{\tau'}) \tag{6.27}$$

In this case the identification of $\tilde{Y}$ with a commutative scheme is given by β. Thus if $\mathcal{T}$ is a quasicoherent $\mathcal{O}_Y$-module then

$$\mathcal{M} \otimes_{o_{\tilde{Y}}} \beta^*(\mathcal{T}) = \beta^*(\beta_*\mathcal{M} \otimes_{o_Y} \mathcal{T})$$

We use this in the computation below.

$$\begin{aligned}
\mathcal{M} \otimes_{o_{\tilde{Y}}} (\beta^*(m_{Y,p}\mathcal{N}))_{\tau'} &= \tau'_*(\mathcal{M} \otimes_{o_{\tilde{Y}}} \beta^*(m_{Y,p}\mathcal{N})) \qquad \text{(See (5.2))}\\
&= (\beta^{-1})_*\tau_*\beta_*\beta^*(\beta_*\mathcal{M} \otimes_{o_Y} m_{Y,p}\mathcal{N})\\
&= (\beta^{-1})_*\tau_*(\beta_*\mathcal{M} \otimes_{o_Y} m_{Y,p}\mathcal{N})
\end{aligned}$$

Thus, using (6.24), (6.27) reduces to

$$\beta_*(\mathcal{M}(1)) = \tau_*(\beta_*\mathcal{M} \otimes_{o_Y} m_{Y,p}\mathcal{N})$$

Now the righthand side of this equation is equal to

$$\beta_*\mathcal{M} \otimes_{o_Y} m_{Y,p}\mathcal{N}_\tau = \beta_*\mathcal{M} \otimes_{o_Y} I_Y$$

So finally we have to show

$$\beta_*(\mathcal{M}(1)) = \beta_*\mathcal{M} \otimes_{o_Y} I_Y$$

But this follows easily from (6.21).

p is a fixed point for τ. Now we identify $\tilde{Y}$ with the ordinary commutative blowup of Y at p. Denote the latter by $\tilde{Y}'$. We have $\tilde{Y}' = \operatorname{Proj} \mathcal{E}_Y$ where $\mathcal{E}_Y$ is as in the proof of Theorem 6.2.2. Let $\beta' : \tilde{Y}' \to Y$ be the structure morphism. There is now a commutative diagram

$$\begin{array}{ccc} \tilde{Y} & \xrightarrow{\gamma} & \tilde{Y}' \\ {\scriptstyle\beta}\big\downarrow & & {\scriptstyle\beta'}\big\downarrow \\ Y & = & Y \end{array}$$

where $\gamma : \tilde{Y} \to \tilde{Y}'$ denotes the identification. Note however that γ does not commute with the canonical shift functors. Translating (6.17) we find that

$$\gamma_*(\mathcal{M}(1)) = (\gamma_*\mathcal{M})(1) \otimes_{o_Y} o_Y(Y) \tag{6.28}$$

If we now look back at the definition of τ' then we see that it was in fact defined on $\tilde{Y}'$ and then pulled back to $\tilde{Y}$ by γ. Thus

$$\tau' = \gamma^{-1}\tau''\gamma$$

where τ'' is the extension of τ to $\tilde{Y}'$.

We compute

$$\mathcal{M} \otimes_{o_{\tilde{Y}}} \beta^{-1}(m^{\mu}_{Y,p}\mathcal{N})_{\tau'} = (\gamma^{-1})_*\tau''_*(\gamma_*\mathcal{M} \otimes_{o_{\tilde{Y}'}} \beta'^{-1} m^{\mu}_{Y,p}\mathcal{N})$$

This equality is a formal computation, analogous to the one where p is not a fixed point.

Using (6.24) we now have to show

$$\gamma_*(\mathcal{M}(\mu) \otimes_{o_Y} o_Y((1-\mu)Y)) = \tau''_*(\gamma_*\mathcal{M} \otimes_{o_{\tilde{Y}'}} \beta'^{-1}(m^{\mu}_{Y,p}\mathcal{N})) \tag{6.29}$$

From (6.28) it follows that the lefthand side of this equation is equal to

$$\gamma_*\mathcal{M}(\mu) \otimes_{o_Y} o_Y(Y) = (\gamma_*\mathcal{M} \otimes_{o_Y} o_Y(Y))(\mu) \tag{6.30}$$

Now we translate everything to graded modules. Let $\gamma_*\mathcal{M}$ be represented by $\mathcal{P}$. Then the righthand side of (6.29) is represented by

$$\tau_*(\mathcal{P} \otimes_{\mathcal{E}_Y} (m^{\mu}_{Y,p}\mathcal{E}_Y \otimes_{o_Y} \mathcal{N}))$$

(6.30) is represented by

$$\tau_*(\mathcal{P} \otimes_{o_Y} \mathcal{N})(\mu) = \tau_*(\mathcal{P} \otimes_{\mathcal{E}_Y} (\mathcal{E}_Y(\mu) \otimes_{o_Y} \mathcal{N}))$$

Thus it is sufficient to show that up to right bounded bimodules

$$m^{\mu}_{Y,p}\mathcal{E}_Y = \mathcal{E}_Y(\mu)$$

This is now clear.

6.5. Birationality

Here the notations are as in the previous section. We will show that X and $\tilde{X}$ are isomorphic "outside the τ-orbit of p". In particular we may view X and $\tilde{X}$ as being birational.

Define $\alpha^{-1}(\mathcal{C}_p)$ as in §3.11. Thus the objects in $\alpha^{-1}(\mathcal{C}_p)$ are the objects in $\operatorname{Mod}(\tilde{X})$ that are represented by graded $\mathcal{D}$-modules such that $\mathcal{M}_n \in \mathcal{C}_p$ for all n. $\alpha^{-1}(\mathcal{C}_p)$ has the following slightly more intrinsic description.

LEMMA 6.5.1. *$\alpha^{-1}(\mathcal{C}_p)$ is the full subcategory of objects $\mathcal{M}$ in $\operatorname{Mod}(\tilde{X})$ for which one has $\alpha_*\mathcal{M}(n\tilde{Y}) \in \mathcal{C}_p$ for all n.*

PROOF. From (3.23) together with (6.24) it easily follows that $\alpha_*\mathcal{M}(n\tilde{Y}) \in \mathcal{C}_p$ if $\mathcal{M} \in \alpha^{-1}(\mathcal{C}_p)$.

Let us now prove the converse inclusion. Assume that $\alpha_*\mathcal{M}(n\tilde{Y}) \in \mathcal{C}_p$ for all n. We have to show that $\mathcal{M} \in \alpha^{-1}(\mathcal{C}_p)$. Let $\mathcal{M} = \pi\mathcal{N}$ for $\mathcal{N} \in \mathrm{Gr}(\mathcal{D})$. It suffices to consider the case that $\mathcal{N}$ is noetherian. We will prove that $\mathcal{N}_n \in \mathcal{C}_p$ for $n \gg 0$. This implies $\mathcal{M} \in \alpha^{-1}(\mathcal{C}_p)$.

According to (6.24)

$$\mathcal{M}(n\tilde{Y}) = \pi\mathcal{N}(n\mu) \otimes_{o_X} o_X(n(1-\mu)Y) \tag{6.31}$$

So if $\mu = 1$ then this formula says that $\tilde{\mathcal{N}}_n \in \mathcal{C}_p$ for all n. Since $\mathcal{D}$ satisfies χ (Theorem 6.2.2) it follows from lemma 3.8.3 that $\tilde{\mathcal{N}}_n = \mathcal{N}_n$ for $n \gg 0$. Hence we are done.

This reasoning still works if $\mu > 1$ since $- \otimes_{o_X} o_X(Y)$ commutes with α_*. Hence (6.31) yields that $\tilde{\mathcal{N}}_{n\mu} \in \mathcal{C}_p$ for all n. Then according to lemma 3.12.1 we have modulo $\mathrm{Tors}(\mathcal{D})$

$$\mathcal{N} = \tilde{\mathcal{N}} = \tilde{\mathcal{N}}^{(\mu)} \otimes_{\mathcal{D}^{(\mu)}} \mathcal{D}$$

Let $\mathcal{N}'$ be the module on the righthand side of this equation. It is clear that $\mathcal{N}'_n \in \mathcal{C}_p$ for all n. Since $\mathcal{N}$ and $\mathcal{N}'$ are isomorphic in $\mathrm{QGr}(\mathcal{D})$ and $\mathcal{N}$ is noetherian, it easily follows that $\mathcal{N}_n \in \mathcal{C}_p$ for $n \gg 0$. □

PROPOSITION 6.5.2. *We have*

1. α^* *and* α_* *define inverse equivalences between the categories* $\mathrm{Mod}(X)/\mathcal{C}_p$ *and* $\mathrm{Mod}(\tilde{X})/\alpha^{-1}(\mathcal{C}_p)$.
2. *The functor* α_* *sends* $\alpha^{-1}(\mathcal{C}_p)$ *to* $\mathcal{C}_p$ *and the functor* $R^1\alpha_*$ *sends* $\mathrm{Mod}(\tilde{X})$ *to* $\mathcal{C}_p$.
3. *The functor* α^* *sends* $\mathcal{C}_p$ *to* $\alpha^{-1}(\mathcal{C}_p)$ *and the functor* $L_1\alpha^*$ *sends* $\mathrm{Mod}(X)$ *to* $\alpha^{-1}(\mathcal{C}_p)$.
4. *The functors* $R^i\alpha_*$ *and* $L_i\alpha^*$ *preserve coherent objects.*

PROOF.
1. It is clear from Proposition 6.1.1 and Theorem 5.5.10.1 that $\tilde{\mathcal{D}}$ is strongly graded with respect to $\mathcal{C}_p$. What we have to prove now follows from Proposition 3.11.4.
2. The statement about α_* follows from (3.23). The statement about $R^1\alpha_*$ follows from applying α_* to an injective resolution and using the fact that α_* is exact modulo $\mathcal{C}_p$ (by Proposition 3.11.4).
3. This follows from the definition of $L_i\alpha^*$ together with Propositions 6.1.1 and 6.1.2.
4. This follows from Theorem 6.3.1.6 and lemma 3.1.18. □

COROLLARY 6.5.3. *Assume that* $q \in Y$ *is not contained in the* τ*-orbit of* p. *Let* q' *be the unique point of* $\tilde{Y}$ *such that* $\beta(q') = q$. *Then* α^* *and* α_* *define inverse equivalences between* $\mathcal{C}_q$ *and* $\mathcal{C}_{q'}$.

PROOF. Using the foregoing proposition it is sufficient to prove the following

1. $\mathcal{C}_q \cap \mathcal{C}_p = 0$.
2. $\mathcal{C}_{q'} \cap \alpha^{-1}(\mathcal{C}_p) = 0$.
3. $\alpha_*(\mathcal{C}_{q'}) \subset \mathcal{C}_q$.
4. $\alpha^*(\mathcal{C}_q) \subset \mathcal{C}_{q'}$.

1. is clear. To prove 3. it is sufficient to show that $\alpha_* \mathcal{O}_{\tau'^n q'} \in \mathcal{C}_q$ where τ' is as in §6.4. This follows from the fact that this is obviously true for β by (6.25).

Now we prove 2. Suppose $\mathcal{M} \in \mathcal{C}_{q'} \cap \alpha^{-1}(\mathcal{C}_p)$. Then $\alpha_* \mathcal{M}(n\tilde{Y}) \in \mathcal{C}_q \cap \mathcal{C}_p = 0$. From the relative ampleness of $o_{\tilde{X}}(n\tilde{Y})$ we deduce that $\mathcal{M} = 0$.

Finally we prove 4. It is sufficient to show that

$$\alpha^* \mathcal{O}_{\tau^n q} = \beta^* \mathcal{O}_{\tau^n q} \tag{6.32}$$

First note that if $\mathcal{N} \in \mathcal{C}_{f,q}$ then $\hat{\mathcal{N}}_p = 0$. Hence if $\mathcal{S} \in \tilde{\mathcal{C}}_{f,p}$ then by (5.33) we have $\mathcal{T}or_i^{o_X}(\mathcal{N}, \mathcal{S}) = 0$. We deduce that

$$\mathcal{O}_{\tau^n q} \otimes_{o_X} I^n = \mathcal{O}_{\tau^n q} \otimes_{o_X} o_X(nY) = \mathcal{O}_{\tau^n q} \otimes_{o_Y} o_Y(nY) = \mathcal{O}_{\tau^n q} \otimes_{o_Y} I_Y^n$$

We obtain (6.32) by summing over all n and applying π. □

6.6. The exceptional curve

In this section we will for simplicity make use of the object $\mathcal{O}_X$ and consequently of the functor $(-)_{o_X}$ which goes from bimodules on X to objects in $\operatorname{Mod}(X)$ (cfr. §3.5).

From the compatibility of completion with Hom and tensor product we deduce that the following functors

$$\tilde{\mathcal{C}}_{f,p} \xrightarrow{(-)_{o_X}} \mathcal{C}_{f,p} \xrightarrow{\Gamma(X,-)} \operatorname{mod}(k)$$

are faithful and exact. On o_p they act by $o_p \mapsto \mathcal{O}_p \mapsto k$.

In the sequel we will write $\mathcal{O}_L = \alpha^* \mathcal{O}_{\tau p}$. Thus $\mathcal{O}_L = \pi((\mathcal{D}/m_{\tau p}\mathcal{D})_{o_X})$.

The following lemma gives a description of $\mathcal{D}/m_{\tau p}\mathcal{D}$ as $o_X - \mathcal{D}$-bimodule.

LEMMA 6.6.1. *There is an exact sequence of $o_X - \tilde{\mathcal{D}}$-bimodules.*

$$0 \to (o_X(-Y) \otimes_{o_X} \tilde{\mathcal{D}})(1) \to \tilde{\mathcal{D}} \to \mathcal{D}/m_{\tau p}\mathcal{D} \to 0 \tag{6.33}$$

PROOF. From Proposition 6.1.3 (and its proof) it follows easily that

$$o_X(-Y) \otimes_{o_X} I^{n+1} = m_{\tau p} I^n$$

This yields (6.33) by taking direct sums over n. □

It was observed by Smith and Zhang [**30**] that it is possible to develop a formalism such that $\mathcal{O}_L$ is really the structure sheaf of a "non-commutative curve" L. We will say more about this in §6.7.

In the special case that $p = \tau p$ one has that $m_{\tau p}\mathcal{D}$ is a twosided ideal. Then we can simply define $L = \operatorname{Proj} \mathcal{D}/m_p\mathcal{D}$. In this case we will of course use the notation o_L for the algebra on $\tilde{X}$ corresponding to the identity functor on $\operatorname{Mod}(L)$.

We find

PROPOSITION 6.6.2. *Assume that $p = \tau p$. Then*

1. $L \cong \mathbb{P}^1$.
2. *L is embedded as a divisor in $\tilde{X}$.*
3. *There is a commutative diagram*

$$\begin{array}{ccc} L & \xrightarrow{\gamma} & p \\ {\scriptstyle u}\downarrow & & \downarrow{\scriptstyle v} \\ \tilde{X} & \xrightarrow{\alpha} & X \end{array} \tag{6.34}$$

where u, v are the inclusions and γ is isomorphic to the structure morphism $\mathbb{P}^1 \to \operatorname{Spec} k$.

PROOF. 1. Using Proposition 5.6.2 we see that $\operatorname{Gr}(\mathcal{D}/m_p\mathcal{D})$ is equivalent to $\operatorname{Gr}((\mathcal{D}/m_p\mathcal{D})^\wedge)$ and a similar result for QGr. Using the compatibility results for $\widehat{(-)}$ we find

$$(\mathcal{D}/m_p\mathcal{D})^\wedge = R/m \oplus (m/m^2)_{\phi^{-1}} \oplus (m^2/m^3)_{\phi^{-2}} \oplus \cdots \tag{6.35}$$

where R is as in Theorem 5.1.4 and m is the maximal ideal of R. It is easily seen that the ring on the right of (6.35) is a noetherian two generator quadratic algebra with one relation. It is well known that the Proj of such a graded ring is $\mathbb{P}^1$.

2. From (6.33) it follows that up to right bounded bimodules we have

$$m_p\mathcal{D} = \mathcal{D}(1)(-Y)$$

Hence, up to right bounded bimodules, $m_p\mathcal{D}$ is an invertible ideal and in particular L is a divisor in $\tilde{X}$.

3. This is a translation of the commutative diagram of o_X-algebras, given by

$$\begin{array}{ccc} \mathcal{D}/m_p\mathcal{D} & \longleftarrow & o_X/m_p \\ \uparrow & & \uparrow \\ \mathcal{D} & \longleftarrow & o_X \end{array} \qquad \square$$

We are now in a position to prove the following result.

THEOREM 6.6.3. *Assume $q \in \tilde{Y}$ if $\tau p \neq p$ and $q \in \tilde{Y}$ or $q \in L$ if $\tau p = p$ (we use the identification $L \cong \mathbb{P}^1$ furnished by Proposition 6.6.2). Then $\mathcal{O}_q$ has finite injective dimension in* $\operatorname{Mod}(\tilde{X})$.

PROOF. First note that $q \to \mathcal{O}_q$ is a one-one correspondence between the points on $\tilde{Y}$ and the simple objects in $\operatorname{Mod}(\tilde{Y})$. Similarly for L and $\operatorname{Mod}(L)$. Therefore we interprete $L \cap \tilde{Y}$ set theoretically as those q such that $\mathcal{O}_q$ lies both in $\operatorname{Mod}(\tilde{Y})$ and in $\operatorname{Mod}(L)$. Put

$$r = \begin{cases} \beta(q) & \text{if } q \in \tilde{Y} \\ p & \text{if } q \in L \end{cases}$$

Using the diagrams (6.34) and (6.20) we see that in all cases $\mathcal{O}_r = \alpha_* \mathcal{O}_q$. Thus r is well defined.

Consider first the case $\tau p \neq p$. Then β is an isomorphism. If $r \notin O_\tau(p)$ then according to Corollary 6.5.3 $\mathcal{C}_r$ is equivalent to $\mathcal{C}_q$. Since both these categories are stable under injective hulls (Proposition 5.1.3) we are through in this case.

However if $r \in O_\tau(p)$ then according to Theorem 5.1.4 r is smooth on Y. Thus q is also smooth on $\tilde{Y}$ hence we can apply lemma 5.1.1.

Consider now the case $\tau p = p$. If $p \in L$ then we can apply lemma 5.1.1 again with L as our curve. So assume now that $q \in \tilde{Y} - \tilde{Y} \cap L$. We then claim that $r \neq p$. Suppose the contrary. Thus there is an isomorphism $\zeta : \mathcal{O}_p \to \alpha_* \mathcal{O}_q$. By adjointness we obtain a map $\eta : \alpha^* \mathcal{O}_p \to \mathcal{O}_q$. This map must be non-zero since otherwise ζ had to be zero also. Since $\mathcal{O}_q$ is simple we obtain that η is an epimorphism.

Now the definition of L implies that $\alpha^*\mathcal{O}_p \in \mathrm{Mod}(L)$. Hence since $\mathrm{Mod}(L)$ is weakly closed in $\mathrm{Mod}(\tilde{X})$ we find $\mathcal{O}_q \in \mathrm{Mod}(L)$. Contradiction.

Since $r \notin O_\tau(p) = \{p\}$ we can now use the same reasoning as in the case $\tau p \neq p$ to conclude that $\mathcal{O}_q$ has finite injective dimension. □

6.7. The structure of the exceptional curve

In this section we prove Proposition 6.7.1 below. It is a generalization of the main result of [**30**] in our special case. Throughout $n = |O_\tau(p)|$.

PROPOSITION 6.7.1. *Let $L = \mathrm{Mod}(L)$ be the full abelian subcategory of $\mathrm{Mod}(\tilde{X})$ whose objects are direct limits of subquotients of finite sums $\mathcal{O}_L(m_1)\oplus\cdots\oplus\mathcal{O}_L(m_n)$ (note that this use of L is consistent with the use in §6.6, when $n = 1$).*

Let S be one of the following graded rings.

1. *If $n = 1$ then S is the twist [**9, 38**] of $\mathrm{gr}_F R$ by the automorphism ϕ, which was introduced in Theorem 5.1.4. Here F denotes the m-adic filtration on R.*
2. *If $2 \leq n < \infty$ then $S = k[u,v]$ where $\deg u = 1$, $\deg v = n$.*
3. *If $n = \infty$ then $S = k[x]$ with $\deg x = 1$.*

Then

$$L \cong \begin{cases} \mathrm{QGr}\, S & \text{if } n < \infty \\ \mathrm{Gr}\, S & \text{if } n = \infty \end{cases}$$

This equivalence is compatible with the natural shift functors and sends $\mathcal{O}_L$ to S.

The proof of this result depends on n. If $n = \infty$ then the result follows from [**30**], so we treat this case first.

The case $n = \infty$ In this case we have an exact sequence (by (6.24))

$$0 \to \mathcal{O}_L(-1) \to \mathcal{O}_L \to \mathcal{O}_{L,\tilde{Y}} \to 0$$

and

$$\mathcal{O}_{L,\tilde{Y}} = j^*\alpha^*\mathcal{O}_{\tau p} = \mathcal{O}_{\tau\beta^{-1}(p)} \tag{6.36}$$

(where we use $\alpha j = i\beta$). Proposition 6.7.1 follows from [**30**] if we can show that $\mathcal{O}_{\tau\beta^{-1}(p)}$ is the only simple quotient of $\mathcal{O}_L$.

Hence let $\mathcal{O}_L \to \mathcal{S}$ be such a simple quotient. Tensoring with $\mathcal{O}_{\tilde{Y}}$ we find that $\mathcal{S}_Y$ is a quotient of $\mathcal{O}_{\tau\beta^{-1}(p)}$. Since $\mathcal{O}_{\tau\beta^{-1}(p)}$ is simple, we either have $\mathcal{S}_{\tilde{Y}} = \mathcal{O}_{\tau\beta^{-1}(p)}$ or $\mathcal{S}_{\tilde{Y}} = 0$. If we are in the first case then $\mathcal{O}_{\tau\beta^{-1}(p)}$ is a quotient of $\mathcal{S}$ and hence by the simplicity of $\mathcal{S}$: $\mathcal{S} = \mathcal{O}_{\tau\beta^{-1}(p)}$. Hence assume that we are in the second case. Let $\mathcal{T}$ be some graded quotient of $(\mathcal{D}/m_{\tau p}\mathcal{D})_{o_X}$ such that $\mathcal{S} = \pi\mathcal{T}$, Then $0 = \mathcal{S}_{\tilde{Y}} = \mathcal{S}/\mathcal{S}(-Y) = \mathcal{S}/\mathcal{S}(-1)$ implies that $\mathcal{T}_m \cong \mathcal{T}_{m+1}$ as o_X-modules for $m \gg 0$.

From the next lemma it follows that for $m \gg 0$, $\mathcal{T}_m$ has a composition series starting with $(\mathcal{O}_{\tau p}(mY), \mathcal{O}_{\tau p}((m-1)Y), \dots)$, which is clearly incompatible with the isomorphism $\mathcal{T}_m \cong \mathcal{T}_{m+1}$. Hence we have obtained a contradiction. We conclude that $\mathcal{S} = \mathcal{O}_{\tau\beta^{-1}(p)}$ and so we can invoke the results of [**30**].

The following lemma was used.

LEMMA 6.7.2. *Fix $t \in \mathbb{N}$. Then $(\mathcal{D}/m_{\tau p}\mathcal{D})_{o_X,t}$ is uniserial of length $t+1$ with composition series. $(\mathcal{O}_{\tau p}(tY), \dots, \mathcal{O}_{\tau p}(Y), \mathcal{O}_{\tau p})$ (starting from the top).*

PROOF. By definition $(\mathcal{D}/m_{\tau p}\mathcal{D})_t$ is given by

$$(m_p \cdots m_{\tau^{1-t}p})/(m_{\tau p} m_p \cdots m_{\tau^{1-t}p})$$

We can compute this by completing (for example using Proposition 5.2.2). We find

$$(\mathcal{D}/m_{\tau p}\mathcal{D})_t^{\wedge} \cong (\dots, 0, R/m, \dots, R/m, 0, \dots, \dots)$$

where the R/m occur in positions $1, \dots, 1+t$. It is easy to see that this is a uniserial right C_p-module with the correct composition series. $\square$

The case $n = 1$ This follows from the proof of Proposition 6.6.2.1.

The case $2 \leq n < \infty$ From the viewpoint of computations this is the most interesting case.

According to Proposition 5.6.2 we may clearly assume that $X = \operatorname{Spec} C_p$, $Y = \operatorname{Spec} C_p/(N)$. Note that one has $C_p/(N) = R/(U) \oplus \cdots \oplus R/(U)$.

We should now analyze the blowup $\tilde{X}$ of X at the point defined by $m_0 \subset C_p$ (which according to our current conventions corresponds to p). However it turns out that it is slightly more convenient to work with m_{n-1}. This does not alter our results in any way since all maximal ideals in C_p are conjugate under the automorphism induced by N.

By definition

$$\begin{aligned}\mathcal{D} &= C_p \oplus m_{n-1}N^{-1} \oplus (m_{n-1}N^{-1})^2 \oplus \cdots \\ &= C_p \oplus m_{n-1}N^{-1} \oplus m_{n-1}m_{n-2}N^{-2} \oplus \cdots\end{aligned}$$

The n'th Veronese of $\mathcal{D}$ is given by

$$\mathcal{D}^{(n)} = C_0 \oplus JN^{-n} \oplus J^2 N^{-2n} \oplus \cdots$$

where $J = m_{n-1}m_{n-2}\cdots m_0$. Since $N^n J N^{-n} = J$ we find that conjugation by N^n induces an automorphism of $\mathcal{D}^{(n)}$. Twisting by this automorphism [**9, 38**] yields that $\operatorname{Gr}(\mathcal{D}^{(n)})$ is equivalent to $\operatorname{Gr}(\mathcal{U})$ where $\mathcal{U}$ is the ordinary Rees algebra of the ideal $J \subset C_p$.

Now before we continue, we remind the reader that taking Veronese's and twisting is in general *not* compatible with the natural shift-functors. However here we have $o_{\tilde{X}}(1) = o_{\tilde{X}}(\tilde{Y})$. That is, the shift functor on $\tilde{X}$ is defined by a divisor. Hence to keep track of this shift functor, we simply have to keep track of $\tilde{Y}$. As we will see this is easy.

To make progress we have to compute J explicitly. Using the material in Proposition 5.2.2 or directly we find that

$$J = \begin{pmatrix} m & (U) & \cdots & \cdots & (U) \\ \vdots & \ddots & \ddots & & \vdots \\ \vdots & & \ddots & \ddots & \vdots \\ \vdots & & & \ddots & (U) \\ m & \cdots & \cdots & \cdots & m \end{pmatrix}$$

Computing the powers of J yields $J^n = m^{n-1}J$.

Let T be the Reesring of R associated to m. We find.

$$\mathcal{U} = \begin{pmatrix} T & RU \oplus TU(-1) & \cdots & \cdots & RU \oplus TU(-1) \\ \vdots & \ddots & \ddots & & \vdots \\ \vdots & & \ddots & \ddots & \vdots \\ \vdots & & & \ddots & RU \oplus TU(-1) \\ T & \cdots & \cdots & \cdots & T \end{pmatrix}$$

In particular we have an inclusion

$$\begin{pmatrix} T & TU(-1) & \cdots & \cdots & TU(-1) \\ \vdots & \ddots & \ddots & & \vdots \\ \vdots & & \ddots & \ddots & \vdots \\ \vdots & & & \ddots & TU(-1) \\ T & \cdots & \cdots & \cdots & T \end{pmatrix} \hookrightarrow \mathcal{U} \tag{6.37}$$

with rightbounded cokernel. Thus $\tilde{X} = \operatorname{Proj} \mathcal{D} = \operatorname{Proj} \mathcal{D}^{(n)} \cong \operatorname{Proj} \mathcal{U}$ is equal to the Proj of the lefthand side of (6.37).

There is a more elegant way to look at this. Let $X_c = \operatorname{Spec} R$, $Y_c = \operatorname{Spec} R/(U)$, $\tilde{X}_c = \operatorname{Proj} T$. Let T_Y be the Reesring of $R/(U)$ and put $\tilde{Y}_c = \operatorname{Proj} Y_c$. Define $\mathcal{A}$ to be the following algebra on $\tilde{X}_c$:

$$\mathcal{A} = \begin{pmatrix} o_{\tilde{X}_c} & o_{\tilde{X}_c}(-\tilde{Y}_c) & \cdots & \cdots & o_{\tilde{X}_c}(-\tilde{Y}_c) \\ \vdots & \ddots & \ddots & & \vdots \\ \vdots & & \ddots & \ddots & \vdots \\ \vdots & & & \ddots & o_{\tilde{X}_c}(-\tilde{Y}_c) \\ o_{\tilde{X}_c} & \cdots & \cdots & \cdots & o_{\tilde{X}_c} \end{pmatrix}$$

Then we obtain $\tilde{X} = \operatorname{Spec} \mathcal{A}$ in Qsch/X_c. A similar computation yields that $\tilde{Y} = \operatorname{Spec} \mathcal{C}$ where $\mathcal{C} = \operatorname{diag}(o_{\tilde{Y}_c}, \dots, o_{\tilde{Y}_c})$. Furthermore $o_{\tilde{X}}(-\tilde{Y})$ is given by the twosided ideal

$$\mathcal{J} = \begin{pmatrix} o_{\tilde{X}_c}(-\tilde{Y}_c) & o_{\tilde{X}_c}(-\tilde{Y}_c) & \cdots & \cdots & o_{\tilde{X}_c}(-\tilde{Y}_c) \\ \vdots & \ddots & \ddots & & \vdots \\ \vdots & & \ddots & \ddots & \vdots \\ \vdots & & & \ddots & o_{\tilde{X}_c}(-\tilde{Y}_c) \\ o_{\tilde{X}_c} & \cdots & \cdots & \cdots & o_{\tilde{X}_c}(-\tilde{Y}_c) \end{pmatrix} \tag{6.38}$$

in $\mathcal{A}$.

Now we compute the exceptional curve. By definition this will correspond to $\pi(\mathcal{U}/m_0\mathcal{U})$. A quick computation reveals that

$$\mathcal{U}/m_o\mathcal{U} \cong (\operatorname{gr} R \; U\operatorname{gr} R(-1) \; \cdots \; U\operatorname{gr} R(-1))$$

where $\operatorname{gr} R$ is the associated graded ring for the m-adic filtration on R.

Let $\mathcal{O}_{L_c}$ be the exceptional curve in $\tilde{X}_c$. Then we find that the exceptional curve in $\tilde{X}$ is given by

$$\begin{aligned}\mathcal{O}_L &= (\mathcal{O}_{L_c}\ \mathcal{O}_{L_c}(-\tilde{Y}_c)\ \cdots\ \mathcal{O}_{L_c}(-\tilde{Y}_c)) \\ &= (\mathcal{O}_{L_c}\ \mathcal{O}_{L_c}(-p_c')\ \cdots\ \mathcal{O}_{L_c}(-p_c'))\end{aligned} \tag{6.39}$$

where $p_c' = L_c \cap \tilde{Y}_c$.

Tensoring (6.39) with positive and negative powers of $\mathcal{J}$ we find that $\operatorname{Mod}(L)$ is contained in the category of modules over

$$\mathcal{B} = \begin{pmatrix} o_{L_c} & o_{L_c}(-p_c') & \cdots & \cdots & o_{L_c}(-p_c') \\ \vdots & \ddots & \ddots & & \vdots \\ \vdots & & \ddots & \ddots & \vdots \\ \vdots & & & \ddots & o_{L_c}(-p_c') \\ o_{L_c} & \cdots & \cdots & \cdots & o_{L_c} \end{pmatrix}$$

The inherited shift functor on $\operatorname{Mod}(\mathcal{B})$ is given by tensoring with the inverse of

$$\mathcal{I} = \begin{pmatrix} o_{L_c}(-p_c') & o_{L_c}(-p_c') & \cdots & \cdots & o_{L_c}(-p_c') \\ \vdots & \ddots & \ddots & & \vdots \\ \vdots & & \ddots & \ddots & \vdots \\ \vdots & & & \ddots & o_{L_c}(-p_c') \\ o_{L_c} & \cdots & \cdots & \cdots & o_{L_c}(-p_c') \end{pmatrix}$$

To interprete this we have to remember that $L_c \cong \mathbb{P}^1$. So we can view $\mathcal{B}$ as an ordinary sheaf of algebras on $\mathbb{P}^1$ given by

$$\mathcal{B} = \begin{pmatrix} \mathcal{O}_{\mathbb{P}^1} & \mathcal{O}_{\mathbb{P}^1}(-1) & \cdots & \cdots & \mathcal{O}_{\mathbb{P}^1}(-1) \\ \vdots & \ddots & \ddots & & \vdots \\ \vdots & & \ddots & \ddots & \vdots \\ \vdots & & & \ddots & \mathcal{O}_{\mathbb{P}^1}(-1) \\ \mathcal{O}_{\mathbb{P}^1} & \cdots & \cdots & \cdots & \mathcal{O}_{\mathbb{P}^1} \end{pmatrix}$$

and $\mathcal{I}$ as the corresponding sheaf of twosided ideals. With this new point of view, $\mathcal{O}_L$ is given by

$$\mathcal{O}_L = (\mathcal{O}_{\mathbb{P}^1}\ \mathcal{O}_{\mathbb{P}^1}(-1)\ \cdots\ \mathcal{O}_{\mathbb{P}^1}(-1)) \tag{6.40}$$

From this explicit interpretation it is now easy to verify that every object in $\operatorname{Mod}(\mathcal{B})$ is a direct limit of subquotients of direct sums of objects like (6.40), tensored with powers of $\mathcal{I}$. Hence $L \cong \operatorname{Spec} \mathcal{B}$. What remains to be shown is that $\operatorname{Spec} \mathcal{B} = \operatorname{Proj} S$.

One way to accomplish this is by considering the triple [**6**] $(\operatorname{Mod}(\mathcal{B}), \mathcal{O}_L, \mathcal{I}^{-1})$ and to verify that this triple is ample. The corresponding graded ring

$$\Gamma_*(\mathcal{O}_L) = \sum \operatorname{Hom}(\mathcal{O}_L, \mathcal{O}_L \otimes_{\mathcal{B}} \mathcal{I}^{-n})$$

is equal to S. Hence according to [**6**] : $\operatorname{Mod}(\mathcal{B}) = \operatorname{QGr}(\mathcal{B})$. This finishes the proof of Proposition 6.7.1 in all cases.

LEMMA 6.7.3. *If $\mathcal{M} \in \alpha^{-1}(\mathcal{C}_p)$ then there is a non-zero map $\mathcal{O}_L(t) \to \mathcal{M}$ for some t.*

PROOF. Assume that $\mathcal{M} = \pi\mathcal{N}$ where $\mathcal{N} \in \mathrm{Gr}(\mathcal{D})$ is torsionfree (for $\mathrm{Tors}(\mathcal{D})$). Let u be such that $\mathcal{N}_u \neq 0$. Since $\mathcal{N}_u \in \mathcal{C}_p$, $\mathcal{N}_u$ will contain some $\mathcal{O}_{\tau^v p}$. Hence there is a non-zero map

$$(\mathcal{O}_{\tau^v p} \otimes_{ox} \mathcal{D})(-u) \to \mathcal{N}$$

Since $\mathcal{N}$ is torsion free this yields a non-zero map.

$$\alpha^*(\mathcal{O}_{\tau^v p})(-u) \to \mathcal{M} \tag{6.41}$$

If it happens that $\tau^v p = \tau p$ then $\alpha^*(\mathcal{O}_{\tau^v p}) = \mathcal{O}_L$ and we are through.

Assume $\tau^v p \neq \tau p$. Then according to Proposition 8.3.2

$$\alpha^*(\mathcal{O}_{\tau^v p}) = \mathcal{O}_{\tau^v \beta^{-1}(p)} \tag{6.42}$$

There is an epimorphism

$$\mathcal{O}_L = \alpha^*(\mathcal{O}_{\tau p}) \to \beta^*(\mathcal{O}_{\tau p}) = \mathcal{O}_{\tau\beta^{-1}(p)}$$

Twisting yields an epimorphism

$$\mathcal{O}_L(v-1) \to \mathcal{O}_{\tau\beta^{-1}(p)}(v-1) = \mathcal{O}_{\tau^v \beta^{-1}(p)} \tag{6.43}$$

Combining (6.41)(6.42)(6.43) yields what we want. □

COROLLARY 6.7.4. *Every object in* $\alpha^{-1}(\mathcal{C}_p) \cap \mathrm{mod}(\tilde{X})$ *is a finite extension of objects in* $\mathrm{mod}(L)$.

PROOF. This follows immediately from lemma 6.7.3. □

So informally we can say that $\alpha^{-1}(\mathcal{C}_p)$ consists the objects on $\tilde{X}$ supported on L.

6.8. The strict transform

This is a more specialized section. We introduce the notion of a strict transform and its influence on the invariant $T_p(\mathcal{F})$ introduced in §5.7. *As in §5.7 we assume that the* τ*-orbit of* p *has infinite order.*

First note some lemmas.

LEMMA 6.8.1. *One has* $\alpha^{-1}(\mathcal{C}_p) \cap \mathrm{mod}(\tilde{X}) \subset \mathrm{trans}_{\tilde{Y}}(\tilde{X})$.

PROOF. This follows for example from Corollary 6.7.4 together with the fact that up to finite length modules every object in $\mathrm{mod}(L)$ is isomorphic to $\mathcal{O}_L^t$ for some t (see [**30**] or Proposition 6.7.1). It then suffices to invoke (6.36). □

LEMMA 6.8.2. *The functors* α^*, α_*, *restrict to functors between* $\mathrm{trans}_Y(X)$ *and* $\mathrm{trans}_{\tilde{Y}}(\tilde{X})$.

PROOF. That α^* preserves transversality follows immediately from $\alpha j = i\beta$.

Let us now look at α_*. Take an object $\mathcal{T} \in \mathrm{trans}_{\tilde{Y}}(\tilde{X})$. We have to show that $i^*\alpha_*\mathcal{T}$ has finite length. For this it is sufficient to show that $\beta^* i^* \alpha_* \mathcal{T} = j^*\alpha^*\alpha_*\mathcal{T}$ has finite length. Now $\alpha^*\alpha_*\mathcal{T}$ is isomorphic to $\mathcal{T}$ modulo $\alpha^{-1}(\mathcal{C}_p) \cap \mathrm{mod}(\tilde{X})$ (Theorem 6.5.2) so it is in $\mathrm{trans}_{\tilde{Y}}(\tilde{X})$ by lemma 6.8.1. This proves what we want. □

If $\mathcal{F} \in \mathrm{Mod}(X)$ then we define

$$\begin{aligned}\mathcal{F}I^n &= \mathrm{im}(\mathcal{F} \otimes I^{\otimes n} \to \mathcal{F}(nY)) \\ \mathcal{F}\cdot\mathcal{D} &= \oplus_n \mathcal{F}I^n \\ \alpha^{-1}(\mathcal{F}) &= \pi(\mathcal{F}\cdot\mathcal{D})\end{aligned}$$

We will use related notations such as $\mathcal{F}\cdot\mathcal{D}_Y$, $\beta^{-1}(\mathcal{F})$ when they apply.

The idea is that if $\mathcal{F} \in \mathrm{trans}_Y(X)$ then $\alpha^{-1}(\mathcal{F})$ should correspond to the strict transform of $\mathcal{F}$ on $\tilde{X}$. Unfortunately, unlike in the commutative case $\alpha^{-1}(\mathcal{F})$ depends on $\mathcal{F}$ itself and not only on the image of $\mathcal{F}$ in $\mathrm{mod}(X)/\mathcal{C}_{f,p}$. Therefore we modify this definition as follows.

Definition 6.8.3. Assume that $\mathcal{F} \in \mathrm{trans}_Y(X)$ is Y-torsion free. Then the *strict transform* $\alpha_s^{-1}(\mathcal{F})$ of $\mathcal{F}$ is defined as $(\alpha^{-1}\circ N_p)(\mathcal{F})$ (see §5.7 for the definition of N_p).

Lemma 6.8.4. *$\alpha^{-1}(\mathcal{F})$ and $\alpha_s^{-1}(\mathcal{F})$ are isomorphic to $\alpha^*\mathcal{F}$ modulo $\alpha^{-1}(\mathcal{C}_p)$.*

Proof. This follows from the definition of $\alpha^{-1}(\mathcal{F})$ together with the fact that $\mathcal{T}or_1^{o_X}(\mathcal{F}, o_X(nY)/I^n) \in \mathcal{C}_p$ by Theorem 5.5.10. □

Proposition 6.8.5. *Identify $\tilde{Y}$ and Y via the map β. With this identification we have $R = \hat{\mathcal{O}}_{Y,p} = \hat{\mathcal{O}}_{Y,\beta^{-1}(p)}$. As usual let m be the maximal ideal of R. Assume that $\mathcal{F} \in \mathrm{trans}_Y(X)$ is Y-torsion free. Then we have the following*

1. *$\alpha_s^{-1}(\mathcal{F})$ is $\beta^{-1}(p)$-normalized.*
2. *One has*

$$T_{\beta^{-1}(p)}(\alpha_s^{-1}(\mathcal{F})) = mT_p(\mathcal{F}) \tag{6.44}$$

Proof. Assume that $\mathcal{F}$ is p-normalized.

The first claim is that the exact sequence coming from the inclusion $o_X(-Y) \to o_X$

$$0 \to \mathcal{F}((n-1)Y) \to \mathcal{F}(nY) \to \mathcal{F}_Y(nY) \to 0$$

restricts to an exact sequence (for $n \geq 1$)

$$0 \to \mathcal{F}I^{n-1} \to \mathcal{F}I^n \to \mathcal{F}_Y I_Y^n \to 0$$

Writing out everything this amounts to checking the exactness and well-definedness of the complex

$$0 \to \mathcal{F}m_p \cdots m_{\tau^{-n+2}p}(-Y) \to \mathcal{F}m_p \cdots m_{\tau^{-n+1}p} \to \mathcal{F}_Y m_{Y,p} \cdots m_{Y,\tau^{-n+1}p} \to 0$$

That this is indeed a complex is clear. To show it is exact it suffices to show that

$$\begin{aligned}\mathrm{length}(\mathcal{F}(-Y)/\mathcal{F}m_p \cdots m_{\tau^{-n+2}p}(-Y)) = {}&\mathrm{length}(\mathcal{F}/\mathcal{F}m_p \cdots m_{\tau^{-n+1}p}) \\ &+ \mathrm{length}(\mathcal{F}_Y/\mathcal{F}_Y m_{Y,p} \cdots m_{Y,\tau^{-n+1}p})\end{aligned} \tag{6.45}$$

Since the finite length modules involved all lie in $\mathcal{C}_p$ we can check this by completing. $\mathcal{F}$ is p-normalized so we have $\hat{\mathcal{F}}_{Y,p} = (\cdots 0, T_p(\mathcal{F}), 0 \cdots)$ with the non-zero entry occuring in position 0. From the structure of $\hat{\mathcal{F}}_p$ which is given by lemma 5.7.4. we compute that $(\mathcal{F}m_p \cdots m_{\tau^{-n+1}p})\hat{}_p = \hat{\mathcal{F}}_p m_0 m_{-1} \cdots m_{-n+1}$ is equal to

$$(\cdots T_p(\mathcal{F})\ mT_p(\mathcal{F}) \cdots mT_p(\mathcal{F})\ 0 \cdots 0 \cdots)$$

where now the first $mT_p(\mathcal{F})$ occurs in location $-n+1$ and the first 0 in location 1. In a similar way we find that $(\mathcal{F}_Y m_{Y,p} \cdots m_{Y,\tau^{-n+1}p})\hat{}_p$ is equal to

$$(\cdots 0\ mT_p(\mathcal{F})\ 0 \cdots)$$

Now (5.37) immediately follows.

We conclude that we have a complex of graded $\mathcal{D}$-modules

$$0 \to \mathcal{F} \cdot \mathcal{D}(-1) \to \mathcal{F} \cdot \mathcal{D} \to \mathcal{F}_Y \cdot \mathcal{D}_Y \to \tag{6.46}$$

exact in degree ≥ 1. A simple local computation shows that the canonical map $\mathcal{F}_Y I_Y \otimes_{o_Y} I_Y^{n-1} \to \mathcal{F}_Y I_Y^n$ is an isomorphism and hence, up to right bounded objects, $\mathcal{F}_Y \cdot \mathcal{D}_Y = (\mathcal{F}_Y I_Y \otimes \mathcal{D}_Y)(-1)$.

Applying π to (6.46) we find the exact sequence

$$0 \to \alpha^{-1}(\mathcal{F})(-\tilde{Y}) \to \alpha^{-1}(\mathcal{F}) \to \beta^*(\mathcal{F}_Y I_Y)(-\tilde{Y}) \to 0$$

Now $\mathcal{F}_Y I_Y = \mathcal{F}_Y m_p(Y)$ and using that β is an isomorphism together with the fact that twisting by Y and $\tilde{Y}$ on finite length modules simply amounts to applying τ we obtain $\beta^*(\mathcal{F}_Y I_Y)(-\tilde{Y}) = \beta^*(\mathcal{F}_Y m_p)$. This yields $\alpha^{-1}(\mathcal{F})_{\tilde{Y}} = \beta^*(\mathcal{F}_Y m_p)$ which implies what we have to show. □

PROPOSITION 6.8.6. *Assume that $\mathcal{F} \in \operatorname{trans}_Y(X)$ is Y-torsion free and that q is not in the τ-orbit of p (but that q also has infinite τ-orbit). Then $T_{\beta^{-1}q}(\alpha_s^{-1}(\mathcal{F}))$ is equal to $T_q(\mathcal{F})$.*

PROOF. This is a local verification as in Proposition 6.8.5, but easier. □

We would also like to understand $\alpha^{-1}(\mathcal{F})$ if $\mathcal{F}$ is not normalized. Let us denote the quotient functor $\operatorname{mod}(\tilde{X}) \to \operatorname{mod}(\tilde{X})/\mathcal{C}_f$ by η. The relevant result is the following.

PROPOSITION 6.8.7. *Assume $\mathcal{F} \in \operatorname{trans}_Y(X)$ is Y-torsion free. Then $\eta\alpha_s^{-1}(\mathcal{F})$ is the minimal subobject of $\eta\alpha^{-1}(\mathcal{F})$ such that the quotient lies in the image of $\alpha^{-1}(\mathcal{C}_p)$.*

PROOF. First note that if $\mathcal{F}' \subset \mathcal{F}$ such that $\mathcal{F}/\mathcal{F}' \in \mathcal{C}_p$ then $\alpha^{-1}(\mathcal{F})/\alpha^{-1}(\mathcal{F}') \in \alpha^{-1}(\mathcal{C}_p)$ by lemma 6.8.4 and Proposition 6.5.2.

Let $\mathcal{F}_n$ be the p-normalization of $\mathcal{F}$ (see §5.7). With a similar local verification as in the proof of Proposition 6.8.5 we find that the inclusion $\mathcal{F}_n(-mY) \subset \mathcal{F}_n$, for $m \geq 0$ induces an isomorphism $\alpha^{-1}(\mathcal{F}_n(-mY)) \to \alpha^{-1}(\mathcal{F}_n)$ modulo $\mathcal{C}_f$.

If we take m large enough then we will have an inclusion $\mathcal{F}_n(-mY) \subset \mathcal{F}$. Together with the result of the previous paragraph this yields an inclusion of $\alpha^{-1}(\mathcal{F}_n) \subset \alpha^{-1}(\mathcal{F})$, if we work modulo $\mathcal{C}_f$. Furthermore since $\mathcal{F}_n$ is isomorphic to $\mathcal{F}$ modulo $\mathcal{C}_p$, this inclusion becomes an isomorphism when viewed modulo $\alpha^{-1}(\mathcal{C}_p)$.

Now let us assume that $\mathcal{F}$ is p-normal and let $\mathcal{G}$ be a subobject of $\alpha^{-1}(\mathcal{F})$. Put $\mathcal{H} = \alpha_*(\mathcal{G})$. If we choose m large enough then $\mathcal{F}(-mY) \subset \mathcal{H}$ and hence we have inclusions $\alpha^{-1}\mathcal{F}(-mY) \subset \alpha^{-1}(\mathcal{H}) \subset \mathcal{G} \subset \alpha^{-1}(\mathcal{F})$. Since as above $\alpha^{-1}\mathcal{F}(-mY) = \alpha^{-1}(\mathcal{F})$ modulo $\mathcal{C}_f$, we conclude that $\mathcal{G} = \alpha^{-1}(\mathcal{F})$ modulo $\mathcal{C}_f$. □

If $\mathcal{F} \in \operatorname{trans}_Y(X)$ then let us write $l_p(\mathcal{F})$ for the minimal n such that $m^n T_p(\mathcal{F}) = 0$. We have the following result.

PROPOSITION 6.8.8. *Let $\mathcal{F} \in \operatorname{trans}_{\tilde{Y}}(\tilde{X})$ be $\tilde{Y}$-torsion free. Then $l_p(\alpha_*(\mathcal{F})) \leq l_{\beta^{-1}(p)}(\mathcal{F}) + 1$.*

PROOF. Let $\mathcal{G}$ be the p-normalization of $\alpha_*(\mathcal{F})$. By Proposition 6.8.7 we have modulo $\mathcal{C}_f$ an inclusion $\alpha^{-1}(\mathcal{G}) \subset \alpha^{-1}(\alpha_*(\mathcal{F})) \subset \mathcal{F}$. Hence we have $l_p(\alpha_*(\mathcal{F})) = l_p(\mathcal{G}) \leq l_{\beta^{-1}(p)}(\alpha^{-1}(\mathcal{G})) + 1 \leq l_{\beta^{-1}(p)}(\mathcal{F}) + 1$ (the first inequality follows from (6.44)). □

A similar verification yields

PROPOSITION 6.8.9. *Let $\mathcal{F} \in \operatorname{trans}_{\tilde{Y}}(\tilde{X})$ be $\tilde{Y}$-torsion free and assume that q is not in the τ-orbit of p (but that q also has infinite τ-orbit). Then $T_q(\alpha_*(\mathcal{F})) = T_{\beta^{-1}(q)}(\mathcal{F})$.*

6.9. A result on K_0 of some categories

In this section we use the results on strict transform to prove a technical result which will be used later. *We now assume that the τ-orbit of every point on Y has infinite order.* Thus in particular Y is smoooth.

For a collection of natural numbers $z = (z_o)_{o \in Y/\langle\tau\rangle}$ let us define $\operatorname{trans}_{Y,z}(X)$ as the full subcategory of $\operatorname{trans}_Y(X)$ consisting of objects $\mathcal{F}$ such that $l_q(\mathcal{F}) \leq z_{\bar{q}}$ for $q \in Y$. We define $M_z(X)$ as $\operatorname{trans}_{Y,z}(X)/\mathcal{C}_f$.

For $o \in Y/\langle\tau\rangle$ let us define e_o by $(\delta_{o,o'})_{o' \in Y/\langle\tau\rangle}$. Our aim is to prove the following theorem

THEOREM 6.9.1. *Let $\tilde{X}$ be the blowup of X at p. One has the following relation.*

$$K_0(M_z(\tilde{X})) \cong \begin{cases} K_0(M_{z+e_{\bar{p}}}(X)) \oplus \mathbb{Z} & \text{if } z_{\bar{p}} \geq 1 \\ K_0(M_{z+e_{\bar{p}}}(X)) & \text{if } z_{\bar{p}} = 0 \end{cases}$$

PROOF. By lemma 6.8.1 $\alpha^{-1}(\mathcal{C}_p) \cap \operatorname{mod}(\tilde{X})$ is contained in $\operatorname{trans}_{\tilde{Y}}(\tilde{X})$. Let us write $\mathcal{T} = (\alpha^{-1}(\mathcal{C}_p) \cap \operatorname{mod}(\tilde{X}))/\mathcal{C}_{f,\beta^{-1}(p)} \subset \operatorname{trans}_{\tilde{Y}}(\tilde{X})/\mathcal{C}_f$ and let us temporarily use the notation $\bar{M}_z(\tilde{X}) = M_z(\tilde{X})/(\mathcal{T} \cap M_z(\tilde{X}))$.

We first show that α^{-1}, α_s^* define inverse equivalence between $M_{z+e_{\bar{p}}}(X)$ and $\bar{M}_z(\tilde{X})$. We know already that α^* and α_* define inverse equivalences between $\operatorname{mod}(X)/\mathcal{C}_{f,p}$ and $\operatorname{mod}(\tilde{X})/(\alpha^{-1}(\mathcal{C}_p) \cap \operatorname{mod}(\tilde{X}))$. Since α_s^{-1} and α^* take the same values modulo $\alpha_s^{-1}(\mathcal{C}_p)$, it follows that α_s^{-1} and α_* also define inverse equivalences. Furthermore it follows from Propositions 6.8.5,6.8.6,6.8.8,6.8.9 that α^{-1} and α_* restrict to equivalences between $M_{z+e_{\bar{p}}}(X)$ and $\bar{M}_z(\tilde{X})$.

We now compute the $K_0(\bar{M}_z(\tilde{X}))$ by the localization sequence.

$$K_0(\mathcal{T} \cap M_z(\tilde{X})) \to K_0(M_z(\tilde{X})) \to K_0(\bar{M}_z(\tilde{X})) \to 0 \tag{6.47}$$

If $z_{\bar{p}} = 0$ then $\mathcal{T} \cap M_z(\tilde{X}) = 0$ so in fact $M_z(\tilde{X}) = \bar{M}_z(\tilde{X})$ and we have the corresponding equality on K_0-groups. This proves what we need in the case $z_{\bar{p}} = 0$.

Let us now consider the case $z_{\bar{p}} \geq 1$. In that case $\operatorname{mod}(L)/\mathcal{C}_{f,\beta^{-1}(p)} \subset M_z(\tilde{X})$ and hence the exact sequence (6.47) becomes

$$K_0(\operatorname{mod}(L)/\mathcal{C}_{f,\beta^{-1}(p)}) \xrightarrow{\delta} K_0(M_z(\tilde{X})) \to K_0(\bar{M}_z(\tilde{X})) \to 0$$

where δ is the natural map. By invoking Proposition 6.7.1 one easily deduces that $K_0(\operatorname{mod}(\tilde{X})/\mathcal{C}_{f,\beta^{-1}(p)}) = \mathbb{Z}$, so it remains to show that δ is monic. For $\mathcal{F} \in \operatorname{trans}_{\tilde{Y}}(\tilde{X})$ let $t(\mathcal{F})$ be the degree of $\operatorname{Div}(\mathcal{F})$ (§5.7).

Clearly t is additive on short exact sequences and furthermore it factors through $\operatorname{trans}_{\tilde{Y}}(\tilde{X})/\mathcal{C}_{f,\beta^{-1}(p)}$. The generator of $K_0(\operatorname{mod}(L)/\mathcal{C}_{f,\beta^{-1}(p)})$ is $[\bar{\mathcal{O}}_L]$ and since obviously $t(\bar{\mathcal{O}}_L) = 1$ we conclude that the image of $[\bar{\mathcal{O}}_L]$ cannot be zero under α. Hence α must be monic.

We conclude that if $z_{\bar{p}} \geq 1$ then $K_0(M_z(\tilde{X})) \cong K_0(\bar{M}_z(\tilde{X})) \oplus \mathbb{Z}$. This finishes the proof. $\square$

CHAPTER 7

Derived categories

7.1. Generalities

If Z is a quasi-scheme then in the sequel we use the notation $D^*(Z)$ with $* = \phi, +, -, b$ for the standard derived categories of $\operatorname{Mod}(Z)$. If Z is noetherian then we use $D^*_f(Z)$ for the full subcategories of $D^*(Z)$ whose objects have coherent cohomology.

If $\alpha : Y \to X$ is a morphism between quasi-schemes then since $\operatorname{Mod}(Y)$ has enough injectives it is trivial to define $R\alpha_*$. This is unfortunately not the case for $L\alpha^*$.

We will say that $\operatorname{Mod}(X)$ has enough acyclic objects for α^* if every object in $\operatorname{Mod}(X)$ is a quotient of an object $\mathcal{U}$ in $\operatorname{Mod}(X)$ such that $L_i\alpha^*\mathcal{U} = 0$ for $i > 0$ (cfr §3.9). The following lemma is easy (see [**18**]).

LEMMA 7.1.1. *Assume that* $\operatorname{Mod}(X)$ *has enough acyclic objects for* α^*. *Then* $L\alpha^*$ *exists on* $D^-(X)$ *and can be computed by acyclic resolutions. If* α^* *has finite cohomological dimension then the same is true with* $D(X)$ *replacing* $D^-(X)$.

Assume that $i : Y \to X$ makes Y into a divisor in X (§3.6) and denote by $i^!$ the functor $\mathcal{H}om(o_Y, -)$. From the resolution of o_Y by two invertible bimodules

$$0 \to o_X(-Y) \to o_X \to o_Y \to 0 \tag{7.1}$$

it is clear that $i^!$ has cohomological dimension one. We will need the following lemma.

LEMMA 7.1.2. *Assume that* $\operatorname{Mod}(X)$ *has enough* Y*-torsion free objects. Then for* $\mathcal{F} \in \operatorname{mod}(X)$ *we have*

$$Ri^!\mathcal{F} = (Li^*\mathcal{F})(Y)[-1] \tag{7.2}$$

PROOF. For $\mathcal{F}$ a torsion free object in $\operatorname{Mod}(X)$ this follows immediately by applying $\operatorname{Hom}(-, \mathcal{F})$ to (7.1). The general case follows from [**18**, Prop. 7.4]. □

7.2. Admissible compositions of morphisms between quasi-schemes

In this section we point out a few problems with compositions of morphisms between quasi-schemes which have no equivalent in the commutative case. These problems will be solved in all concrete cases, but they nevertheless represent a nuissance.

Assume that we have a composition of morphisms between quasi-schemes.

$$Z \xrightarrow{\beta} Y \xrightarrow{\alpha} X \tag{7.3}$$

From the very definition of a morphism it is clear that $\alpha_*\beta_*$ and $(\alpha\beta)_*$ are naturally isomorphic (in a canonical way). However in contrast with the commutative case there is no reason why the natural map

$$R(\alpha\beta)_* \to R\alpha_*\, R\beta_* \tag{7.4}$$

should be an isomorphism (for trivial reasons, see the discussion after lemma 7.2.7 below).

This leads to the following definition.

DEFINITION 7.2.1. A composition (α, β) as in (7.3) is *admissible* if (7.4) is an isomorphism (as functors from $D^+(Z)$ to $D^+(X)$).

The following lemma is trivial.

LEMMA 7.2.2. *The composition (α, β) is admissible if and only if for every injective $E \in \operatorname{Mod}(Z)$ one has that $\beta_* E$ is acyclic for α_*.*

For simplicity we will use some variations on Definition 7.2.1. If we work in Qsch/X (i.e. somewhat imprecisely : "if X is a fixed base quasi-scheme"), then we will say that the morphism α is admissible if (7.4) is an isomorphism.

Similarly if $\alpha : (Y, \mathcal{O}_Y) \to (X, \mathcal{O}_X)$ is a morphism of enriched quasi-schemes then we will say that α is admissible if

$$R\Gamma(X, -) \circ R\alpha_* = R\Gamma(Y, -)$$

These conventions are to a certain extent compatible as seen by the following lemma.

LEMMA 7.2.3. *Assume that we have morphisms*

$$Z \xrightarrow{\beta} Y \xrightarrow{\alpha} \operatorname{Spec} R$$

Put $\mathcal{O}_Y = \alpha^* R$, $\mathcal{O}_Z = \beta^* \mathcal{O}_Y$. Then the compositon (α, β) is admissible if and only if the induced map of enriched quasi-schemes $\alpha : (Z, \mathcal{O}_Z) \to (Y, \mathcal{O}_Y)$ is admissible.

Below we will give a few adhoc criteria which will allow us to show that certain morphisms/compositions are admissible. They are all tautologies.

LEMMA 7.2.4. *Assume that we have a composition as in (7.3), but this time assume that we are working in Qsch/W for some base quasi-scheme W. If the morphisms α, β and the composition (α, β) are admissible then so is the morphism $\alpha\beta$.*

LEMMA 7.2.5. *Assume that $\alpha : (Y, \mathcal{O}_Y) \to (X, \mathcal{O}_X)$ is a morphism of enriched quasi-schemes. Then α is admissible if and only if $L_i \alpha^* \mathcal{O}_X = 0$ for $i > 0$.*

COROLLARY 7.2.6. *Let $(X, \mathcal{O}_X)$ be an enriched quasi-scheme. Assume that $\mathcal{A} \in \operatorname{Alg}(X)$. Then $\operatorname{Spec} \mathcal{A} \to X$ is admissible if and only if $\mathcal{T}or^i_{o_X}(\mathcal{O}_X, \mathcal{A})$ is equal to zero for $i > 0$.*

COROLLARY 7.2.7. *Assume that $\alpha : (Y, \mathcal{O}_Y) \to (X, \mathcal{O}_X)$ is a morphism of enriched quasi-schemes which embeds Y as a divisor in X (§3.6). Then α is admisible.*

PROOF. This is a special case of the previous corollary, since $Y = \operatorname{Spec} o_Y$, where we consider o_Y as an o_X algebra. □

This corollary indicates how to make a non-admissible morphism. Take for example commutative schemes Y, X, Y being a Cartier divisor in X and change the structure sheaf of X into one which has Y-torsion.

Below quasi-schemes are often defined as Proj's of algebras. The following lemma is the obvious analogue of Corollary 7.2.6.

LEMMA 7.2.8. *Let $(X, \mathcal{O}_X)$ be an enriched quasi-scheme. Assume that $\mathcal{A}$ is a noetherian graded algebra on X. Then* $\operatorname{Proj}\mathcal{A} \to X$ *is admissible if* $\mathcal{T}or_i^{o_X}(\mathcal{O}_X, \mathcal{A})$ *is right bounded for $i > 0$.*

LEMMA 7.2.9. *Let $f : \mathcal{A} \to \mathcal{B}$ be as in Proposition 3.8.10. Let $\bar{f}$:* $\operatorname{Proj}\mathcal{B} \to \operatorname{Proj}\mathcal{A}$ *be as in Proposition 3.8.11. Then $\bar{f}$ defines an admissible morphism of quasi-schemes in* Qsch/X.

PROOF. This follows from Proposition 3.8.12. □

The following is also standard.

PROPOSITION 7.2.10. *Assume that we have quasi-schemes and morphisms as in* (7.3). *Assume that the composition (α, β) is admissible and that the morphisms α, β, $\alpha\beta$ satisfy the conditions of lemma 7.1.1. Then we have $L(\alpha\beta)^* \cong L\beta^* \, L\alpha^*$ (canonically, as functors from $D^-(X)$ to $D^-(Z)$).*

PROOF. It is easy to see that we have at least a natural transformation

$$L\beta^* \, L\alpha^* \to L(\alpha\beta)^*$$

In fact this can be deduced from the very definition of derived functors (see [**18**, §5]).

Now it is clear that if E runs trough the injective objects in $\operatorname{Mod}(Z)$ then $\operatorname{Hom}_{D(Z)}(-, E)$ is a conservative system of functors on $D(Z)$. Let $\mathcal{M} \in D^-(X)$. Then we have by adjunction and admissibility

$$\begin{aligned}\operatorname{Hom}_{D(Z)}(L\beta^* \, L\alpha^* \mathcal{M}, E) &= \operatorname{Hom}_{D(X)}(\mathcal{M}, R\alpha_* \, R\beta_* E) \\ &= \operatorname{Hom}_{D(X)}(\mathcal{M}, R(\alpha\beta)_* E) \\ &= \operatorname{Hom}_{D(Z)}(L(\alpha\beta)^* \mathcal{M}, E) \quad \square\end{aligned}$$

REMARK 7.2.11. It is easy to see that if we have an admissible composition as in (7.3) then similar results as the ones presented above remain valid on unbounded derived categories provided suitable functors have finite cohomological dimension. In the sequel we will tacitly use such results, leaving the obvious proofs to the reader.

We can now show that some commutative diagrams of quasi-schemes we encountered previously consist of admissible morphisms and admissible compositions.

LEMMA 7.2.12. *All morphisms and compositions of morphisms in the diagram* (6.20) *are admissible.*

PROOF. i and j are admissible because they are divisors (see Theorem 6.3.1.8). β is admissible since it is a map between commutative schemes. To prove that α is admissible it suffices by lemma 7.2.8 to show that $\mathcal{T}or_i^{o_X}(\mathcal{O}_X, \mathcal{D}) = 0$ for $i > 0$. More precisely we have to show that $\mathcal{T}or_i^{o_X}(\mathcal{O}_X, I_p^n) = 0$ for $i > 0$. Since $I_p^n \subset o_X(nY)$ and the quotient is in $\tilde{C}_{f,p}$ this follows from the compatibility of $\mathcal{T}or$ with completion (see Theorem 5.5.10). □

LEMMA 7.2.13. *All morphisms and compositions in diagram* (6.34) *are admissible.*

PROOF. The admissibility of (α, u) follows from lemma 7.2.9. The other compositions and morphisms we leave to the reader. □

CHAPTER 8

The derived category of a non-commutative blowup

8.1. The formalism of semi-orthogonal decompositions

The material in this section is taken from [**14**].

Let $\mathcal{A}$ be a triangulated category and let $\mathcal{B}$, $\mathcal{C}$ be two strict full triangulated subcategories of $\mathcal{A}$. $(\mathcal{B}, \mathcal{C})$ is said to be a *semi-orthogonal pair* if $\operatorname{Hom}_{\mathcal{A}}(B, C) = 0$ for $B \in \mathcal{B}$ and $C \in \mathcal{C}$. Define

$$\mathcal{B}^{\perp} = \{A \in \mathcal{A} \mid \forall B \in \mathcal{B} : \operatorname{Hom}_{\mathcal{A}}(B, A) = 0\}$$

${}^{\perp}\mathcal{C}$ is defined similarly.

The following result is a slight variation of the statement of [**14**, Lemma 3.1].

LEMMA 8.1.1. *The following statements are equivalent for a semi-orthogonal pair $(\mathcal{B}, \mathcal{C})$.*

1. *$\mathcal{B}$ and $\mathcal{C}$ generate $\mathcal{A}$.*
2. *For every $A \in \mathcal{A}$ there exists a distinguished triangle $B \to A \to C$ with $B \in \mathcal{B}$ and $C \in \mathcal{C}$.*
3. *$\mathcal{C} = \mathcal{B}^{\perp}$ and the inclusion functor $i_* : \mathcal{B} \to \mathcal{A}$ has a right adjoint $i^! : \mathcal{A} \to \mathcal{B}$.*
4. *$\mathcal{B} = {}^{\perp}\mathcal{C}$ and the inclusion functor $j_* : \mathcal{C} \to \mathcal{A}$ has a left adjoint $j^* : \mathcal{A} \to \mathcal{C}$*

If one of these conditions holds then the triangles in 2. are unique up to unique isomorphism. They are necessarily of the form

$$i^! A \to A \to j^* A$$

where the morphisms are obtained by adjointness from the identity morphisms $i^! A \to i^! A$ and $j^* A \to j^* A$. In particular triangles as in 2. are functorial.

REMARK 8.1.2. The notations $(i_*, i^!, j_*, j^*)$ are purely symbolic and shouldn't be interpreted as direct and inverse images. In fact in the main application below (Theorem 8.4.1) i_* will be given by an inverse image!

If a pair $(\mathcal{B}, \mathcal{C})$ satisfies one of the conditions of the previous lemma then we say that it is a *semi-orthogonal decomposition* of $\mathcal{A}$. For further reference we note the following diagram of arrows

$$\mathcal{C} \underset{\underset{j^*}{\leftarrow}}{\overset{\overset{j_*}{\rightarrow}}{}} \mathcal{A} \underset{\underset{i_*}{\leftarrow}}{\overset{\overset{i^!}{\rightarrow}}{}} \mathcal{B} \tag{8.1}$$

In the following lemma we give some relations between these arrows.

LEMMA 8.1.3. *One has :*

$$i^! i_* = \mathrm{id}_{\mathcal{B}}$$
$$j^* j_* = \mathrm{id}_{\mathcal{C}}$$
$$j^* i_* = 0$$
$$i^! j_* = 0$$

In the sequel we will slightly extend the meaning of the notion of semi-orthogonality. Assume that we have functors

$$\mathcal{C} \xrightarrow{j_*} \mathcal{A} \xleftarrow{i_*} \mathcal{B}$$

which are fully faithful. Assume furthermore that the essential images of $\mathcal{B}$ and $\mathcal{C}$ in $\mathcal{A}$ are semi-orthogonal in $\mathcal{A}$. Then, if no confusion can arise, we wil also call $(\mathcal{B}, \mathcal{C})$ a semi-orthogonal pair in $\mathcal{A}$. Similarly for a semi-orthogonal decomposition.

Semi-orthogonal decompositions can be constructed starting from a pair of adjoint functors. For an arbitrary functor $F : \mathcal{A} \to \mathcal{B}$ between additive categories let us define $\ker F$ as the full subcategory of $\mathcal{A}$ whose objects satisfy $F(A) = 0$.

LEMMA 8.1.4. *Assume that we have triangulated categories $\mathcal{A}, \mathcal{B}$ and a pair of adjoint functors $i_* : \mathcal{B} \to \mathcal{A}$, $i^! : \mathcal{A} \to \mathcal{B}$ such that $i^! i_* = \mathrm{id}_{\mathcal{B}}$. Then i_* is an embedding of $\mathcal{B}$ in $\mathcal{A}$. The corresponding semi-orthogonal decomposition is given by $(\mathcal{B}, \ker i^!)$.*

8.2. Generalities

In this section the notations and hypotheses of Chapter 6 will be in force. Our aim is to give a non-commutative version of a well-known theorem by Orlov [**27**] which relates $D(\tilde{X})$ to $D(X)$. To this end we need the adjoint functors $R\alpha_*$ and $L\alpha^*$. In particular we need that X has enough acyclic objects for α^* (by lemma 7.1.1).

Therefore at this point we introduce an extra hypothesis which will always hold in the applications.

Let us call an object $\mathcal{L}$ in $\mathrm{mod}(X)$ a *line bundle* around Y if the map $\mathcal{L}(-Y) \to \mathcal{L}$ is injective and if $\mathcal{L}/\mathcal{L}(-Y)$ is a line bundle on Y. Note that $\mathcal{O}_X$ itself is a line bundle on Y. We denote the additive category of objects which are direct sums of line bundles on Y by $\mathcal{V}$.

HYPOTHESIS (**). *Every object in* $\mathrm{Mod}(X)$ *is a quotient of an object in* $\mathcal{V}$.

The following lemma is left to the reader.

LEMMA 8.2.1. *Assume $\mathcal{E} \in \mathcal{V}$.*

1. $\mathcal{T}or_i^{o_X}(\mathcal{E}, o_Y) = 0$ *for* $i \neq 0$.
2. $\mathcal{T}or_i^{o_X}(\mathcal{E}, o_q) = 0$ *for* $i \neq 0$.
3. $\mathcal{T}or_i^{o_X}(\mathcal{E}, \mathcal{D}) = 0$ *for* $i \neq 0$.
4. $\mathcal{T}or_i^{o_X}(\mathcal{E}, \mathcal{D}_Y) = 0$ *for* $i \neq 0$.
5. $L_i\alpha^*\mathcal{E} = 0$ *for* $i \neq 0$.

From this lemma together with Prop. 6.1.2 it is easy to see that $-\overset{L}{\otimes}_{o_X}\mathcal{D}$, $-\overset{L}{\otimes}_{o_X}\mathcal{D}_Y$, $R\alpha^*$ etc ... can be defined in the usual way [**18**] by considering resolutions in $\mathcal{V}$.

For further reference we recall the following.

LEMMA 8.2.2. *$R\alpha_*$ and $L\alpha^*$ have finite cohomological dimension and commute with direct sums.*

PROOF. The fact that $R\alpha_*$ and $L\alpha^*$ have finite cohomological dimension is proved in Theorem 6.3.1.

Since $L\alpha^*$ is the left adjoint to $R\alpha^*$ it is clear that it is compatible with direct sums. Hence let us concentrate on $R\alpha_*$. From the discussion in Section 3.7 it follows that $R\alpha_*$ is equal to $R\omega(-)_0$ and furthermore for $\mathcal{M} \in \mathrm{Gr}(\mathcal{D})$ there is a triangle

$$\begin{array}{ccc} & \mathcal{M} & \\ \swarrow & & \nwarrow \\ R\tau\mathcal{M} & \longrightarrow & R\omega(\pi\mathcal{M}) \end{array} \tag{8.2}$$

Thus it is sufficient to show that $R\tau$ commutes with direct sums. According to Proposition 3.7.4

$$\tau\mathcal{M} = \operatorname{inj}\lim \mathcal{H}om_{\mathcal{D}}(\mathcal{D}/\mathcal{D}_{\geq n}, \mathcal{M})$$

According to Proposition 6.1.1, I_p^n is a coherent $o_X - o_X$ bimodule §3 for all n. Hence this holds also for $\mathcal{D}/\mathcal{D}_{\geq n}$. But then it is easy to see that $\mathcal{D}/\mathcal{D}_{\geq n}$ is also coherent as $\mathcal{D}-\mathcal{D}$-bimodule. Thus $\mathcal{H}om_{\mathcal{D}}(\mathcal{D}/\mathcal{D}_{\geq n}, -)$ commutes with direct sums and hence so does τ. Now τ has finite cohomological dimension (by (8.2) this is the same as α_* having finite cohomological dimension). It is now easy to see that the standard way of defining $R\tau$ on $D(X)$ [**18**] is compatible with direct sums. □

8.3. Computation of some derived functors

We need the following proposition.

PROPOSITION 8.3.1. *Let $n \geq -1$.*

1. $R\alpha_*(\mathcal{O}_L(n)) = (\mathcal{D}/m_{\tau p}\mathcal{D})_{n,o_X}$.
2. *Assume that $\mathcal{E} \in \mathcal{V}$. Then $R\alpha_*((\alpha^*\mathcal{E})(n)) = \mathcal{E} \otimes_{o_X} \tilde{\mathcal{D}}_n$ (cfr. (6.15)).*

PROOF. Since $\mathcal{D}$ satisfies χ (Theorem 6.2.2) it is clear that the proposition is true for $n \gg 0$ (lemma 3.8.3). Our strategy will now be to use descending induction on n. To this end it is convenient to treat the cases $\tau p = p$ and $\tau p \neq p$ separately. In the first case L is divisor in $\tilde{X}$ isomorphic to $\mathbb{P}^1$ so we can use this. In the second case p is smooth on Y (Thm 5.1.4) and so $\beta : \tilde{Y} \to Y$ is an isomorphism. It follows from this that a suitable analog of the proposition is trivially true for Y. We can then lift this to $\tilde{X}$.

In the proof below we will write $\mathcal{E}_{\tau p} = \mathcal{E} \otimes_{o_X} o_{\tau p}$ and similarly $\mathcal{E}_Y = \mathcal{E} \otimes_{o_X} o_Y$.

CASE 1. $\tau p = p$. By admissibility we have $R\alpha_*\mathcal{O}_L(n) = R\gamma_*\mathcal{O}_L(n)$ (where γ is as in diagram (6.34)). Thus 1. follows easily from the corresponding result for $\mathbb{P}^1$.

For 2. we use the exact sequence obtained from (6.33)

$$0 \to \alpha^*(\mathcal{E}(-Y))(1) \to \alpha^*\mathcal{E} \to \alpha^*(\mathcal{E}_{\tau p}) \to 0 \tag{8.3}$$

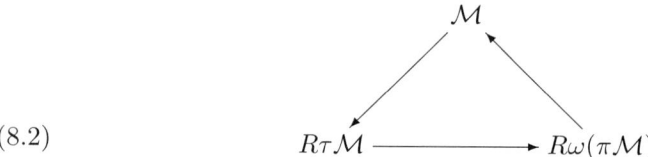

Since $\alpha^*(\mathcal{E}_{\tau p})$ is a direct sum of copies of $\mathcal{O}_L$ we deduce from 1. that

$$R^i\alpha_*(\alpha^*(\mathcal{E}_{\tau p})) = \begin{cases} \mathcal{E}\otimes_{o_X}(\mathcal{D}/m_p\mathcal{D})_n & \text{if } i=0 \\ 0 & \text{otherwise} \end{cases}$$

Assume that 2. is true for a certain $n\geq 0$. Since the autoequivalence $-\otimes_{o_X} o_X(Y)$ is compatible with α_* and α^* (p is a fixed point) 2. will also be true for this n if we replace $\mathcal{E}$ by $\mathcal{E}(mY)$ for arbitrary m.

Tensoring (8.3) with $o_{\tilde{X}}(n-1)$ and applying the long exact sequence for $R\alpha_*$ yields $R^i\alpha_*((\alpha^*\mathcal{E})(n-1))=0$ for $i>0$ and for $i=0$ we obtain an exact sequence

$$0\to\mathcal{E}(-Y)\otimes_{o_X}\mathcal{D}_n\to\alpha_*(\alpha^*(\mathcal{E})(n-1))\to\mathcal{E}\otimes_{o_X}(\mathcal{D}/m_p\mathcal{D})_n\to 0 \tag{8.4}$$

On the other hand if we tensor the exact sequence (6.33) on the left with $\mathcal{E}$ and take the part of degree n then we obtain an exact sequence

$$0\to\mathcal{E}(-Y)\otimes_{o_X}\mathcal{D}_n\to\mathcal{E}\otimes_{o_X}\mathcal{D}_{n-1}\to\mathcal{E}\otimes_{o_X}(\mathcal{D}/m_p\mathcal{D})_n\to 0$$

Comparing with (8.4) using the five lemma completes the proof of the proposition in the case $\tau p = p$.

CASE 2. $\tau p\neq p$. Now p is smooth on Y (Thm. 8.4.1) and the map $\beta:\tilde{Y}\to Y$ is an isomorphism. From (6.16) we obtain an exact sequence

$$0\to\alpha^*(\mathcal{E})(-1)\to\alpha^*(\mathcal{E})\to\beta^*(\mathcal{E}_Y)\to 0 \tag{8.5}$$

By admissibility $R^i\alpha_*((\beta^*\mathcal{E}_Y)(n))=R^i\beta_*((\beta^*\mathcal{E}_Y)(n))$. For $n\geq 0$ we have

$$R^i\beta_*((\beta^*\mathcal{E}_Y)(n)) = \begin{cases} \mathcal{E}_Y\otimes_{o_Y}\mathcal{D}_{Y,n} & \text{if } i=0 \\ 0 & \text{if } i>0 \end{cases}$$

The case $i=0$ follows from (6.21). The case $i>0$ follows from the fact that β is an isomorphism.

Assume that 2. is true for a certain n. Tensor (8.5) with $o_{\tilde{X}}(n)$ and apply $R\alpha_*$. We obtain $R^i\alpha_*((\alpha^*\mathcal{E})(n-1))=0$ for $i\geq 2$ and an exact sequence

(8.6)
$$0\to\alpha_*((\alpha^*\mathcal{E})(n-1))\to\mathcal{E}\otimes_{o_X}\mathcal{D}_n\to\mathcal{E}\otimes_{o_X}\mathcal{D}_{Y,n}\to R^1\alpha_*((\alpha^*\mathcal{E})(n-1))\to 0$$

Tensoring (6.16) on the left with $\mathcal{E}$ and taking the part of degree n yields an exact sequence

$$0\to\mathcal{E}\otimes_{o_X}\mathcal{D}_{n-1}\to\mathcal{E}\otimes_{o_X}\mathcal{D}_n\to\mathcal{E}\otimes_{o_X}\mathcal{D}_{Y,n}\to$$

Comparing with (8.6) using the five lemma yields 2. in this case.

To prove 1. we twist the exact sequence (8.3) by $o_{\tilde{X}}(n)$ and put $\mathcal{E}=\mathcal{O}_X$. Applying $R\alpha_*$ then yields what we want. $\square$

PROPOSITION 8.3.2. *Assume that $q\in Y$. If $q\neq\tau p$ then $L\alpha^*\mathcal{O}_q=\mathcal{O}_{q'}$ with q' such that $\beta(q')=q$ (such a q' is unique!).*

For $q=\tau p$ we have

$$L^i\alpha^*\mathcal{O}_{\tau p} = \begin{cases} \mathcal{O}_L & \text{if } i=0 \\ \mathcal{O}_L(-1) & \text{if } i=1 \\ 0 & \text{otherwise} \end{cases}$$

PROOF. If $q \notin O_\tau(p)$ then the result is an easy excercise, for example using the fact that $I^n \subset o_X(nY)$ with quotient in $\tilde{\mathcal{C}}_{f,p}$ (see Prop. 6.1.1). Hence assume $q \in O_\tau(p)$. We have by lemma 3.9.2

$$L^i \alpha^* \mathcal{O}_q = \pi \, \mathcal{T}or^i_{o_X}(\mathcal{O}_q, \mathcal{D})$$

Now $\mathcal{T}or^i_{o_X}(\mathcal{O}_q, \mathcal{D}) \in \mathcal{C}_p(\mathcal{D})$ by Propositions 6.1.1 and 6.1.2. Since "$\mathcal{T}or$" is compatible with completion (Theorem 5.5.10) it suffices to compute $\operatorname{Tor}^i_{C_p}(\hat{\mathcal{O}}_q, \hat{\mathcal{D}})$. This is a mildly tedious calculation which we leave to the reader. One way to proceed is to use Proposition 5.2.2 to construct a projective resolution of $\hat{\mathcal{O}}_q$ of length 2 over C_p. □

LEMMA 8.3.3. 1. $\operatorname{Hom}_{o_{\tilde{X}}}(\mathcal{O}_L, -)$ *has finite cohomological dimension.*
2. *The functor* $\operatorname{RHom}_{o_{\tilde{X}}}(\mathcal{O}_L, -)$ *is defined on* $D(\tilde{X})$*. For* $\mathcal{M} \in D(X)$ *there is a triangle*

$$R\alpha_*(\mathcal{M}(-1))(Y)$$

(8.7) $$\mathcal{O}_{\tau p} \otimes_k \operatorname{RHom}_{D(\tilde{X})}(\mathcal{O}_L, \mathcal{M}) \longrightarrow R\alpha_* \mathcal{M}$$

3. $\operatorname{RHom}(\mathcal{O}_L, -)$ *commutes with direct sums on* $D(X)$.
4. $\operatorname{RHom}(\mathcal{O}_L, \mathcal{O}_L) = k$.

PROOF. 1. Assume that E is an injective object in $\operatorname{Mod}(\tilde{X})$. We have

$$\mathcal{O}_{\tau p} \otimes_k \operatorname{Hom}_{o_{\tilde{X}}}(\mathcal{O}_L, E) = \mathcal{H}om_{\operatorname{Gr}(\mathcal{D})}(\mathcal{D}/m_{\tau p}\mathcal{D}, \omega E)$$

Now we use the exact sequence (6.33). From exactness of π it follows that ωE is injective. Furthermore assume that $\mathcal{T}$ is a right bounded $o_X - \mathcal{D}$-module. Then adjointness yields that $\operatorname{Hom}_{o_X}(-, \underline{\mathcal{H}om}_{\mathcal{D}}(\mathcal{T}, \omega E))$ vanishes. Thus we deduce that $\underline{\mathcal{H}om}_{\mathcal{D}}(-, \omega E))$ is zero on right bounded $o_X - \mathcal{D}$-bimodules. Hence we obtain from (6.33) an exact sequence in $\operatorname{Mod}(X)$.

$$0 \to \mathcal{H}om_{\operatorname{Gr}(\mathcal{D})}(\mathcal{D}/m_{\tau p}\mathcal{D}, \omega E) \to (\omega E)_0 \to (\omega E)_{-1}(Y) \to 0$$

Which can be rewritten as

(8.8) $$0 \to \mathcal{O}_{\tau p} \otimes_k \operatorname{Hom}_{o_{\tilde{X}}}(\mathcal{O}_L, E) \to \alpha_* E \to \alpha_*(E(-1))(Y) \to 0$$

The fact that α_* has finite cohomogical dimension by 6.3.1 yields what we want.

2. The fact that $\operatorname{R}\mathcal{H}om_{o_{\tilde{X}}}(\mathcal{O}_L, -)$ is defined on the whole of $D(X)$ follows from 1. using [**18**]. Now since α_* also has finite cohomological dimension, every object in $\operatorname{Mod}(\tilde{X})$ has a resolution by objects which are acyclic for both α_* and $\operatorname{Hom}_{o_{\tilde{X}}}(\mathcal{O}_L, -)$ (using the methods of [**18**]). If F is acyclic for $\operatorname{Hom}_{o_{\tilde{X}}}(\mathcal{O}_L, -)$ then it is clear there will be a short exact sequence as in (8.8) with E replaced by F. By considering resolutions with such acyclic objects one finds the triangle (8.7).
3. This follows from 1. together with the construction of $\operatorname{Hom}_{o_{\tilde{X}}}(\mathcal{O}_L, -)$ by acyclic resolutions.
4. This follows from substituting $\mathcal{M} = \mathcal{O}_L$ in the triangle (8.7) and using Proposition 8.3.1. □

The following result is well-known.

LEMMA 8.3.4. *Let $\mathcal{A}$ be an abelian category. Let $X \in D^*(\mathcal{A})$ with $* = +, -$ be an object such that* $\mathrm{Ext}^i(H^j(X), H^{j+1-i}(X)) = 0$ *for $i \geq 2$ and for all j. Then X is isomorphic to the direct sum of its homology groups. The same is true for $X \in D(\mathcal{A})$ if in addition $\mathcal{A}$ satisfies AB4, has enough injectives and* $\mathrm{Hom}(H^j(X), -)$ *has finite cohomological dimension for all j.*

PROOF. Assume first that $X \in D^+(\mathcal{A})$. Write $H(X) = \oplus H^i(X)[-i]$. Note that $H(X)$ is the category theoretic direct sum of the $H^i(X)[-i]$ in $D^+(\mathcal{A})$. We want to construct a quasi-isomorphism $H(X) \to X$. To this end it is sufficient to construct maps $H^i(X)[-i] \to X$ which induce isomorphisms on the i't cohomology. Since $\tau_{\leq i}X \to X$ induces an isomorphism on H^i, it is clearly sufficient to show that the canonical map $\tau_{\leq i}X \to H^i(X)[-i]$ splits. From the triangle

$$\tau_{\leq i-1}X \to \tau_{\leq i}X \to H^i(X)[-i] \to$$

we find that we have to show that

$$\mathrm{Hom}(H^i(X)[-i], \tau_{\leq i-1}X[1]) = 0 \tag{8.9}$$

Now $\tau_{\leq i-1}X$ is a bounded complex and hence (8.9) follows easily by induction from the hypotheses.

The case $X \in D^-(\mathcal{A})$ is similar. Now assume $X \in D(\mathcal{A})$. In this case there are two possible problems with the above reasoning.

1. $\oplus_i H^i(X)[-i]$ is perhaps no longer the direct sum of the $H^i(X)[-i]$ in $D(\mathcal{A})$.
2. $\tau_{\leq i-1}X$ is now an unbounded complex so we can no longer verify (8.9) by induction.

The first difficulty is resolved if we assume that $\mathcal{A}$ satisfies AB4 [**12**]. For the second difficulty we have to show that $\mathrm{Hom}(H^i(X), -)$ is zero on $\mathcal{D}(\mathcal{A})_{\leq -N}$ for $N \gg 0$. Now according to [**18**, Thm 5.1, Cor. 5.3], if $\mathcal{A}$ has enough injectives and $\mathrm{Hom}(H^i(X), -)$ has finite cohomological dimension then we can compute $\mathrm{Hom}(H^i(X), -) = H^0(\mathrm{RHom}(H^i(X), -))$ by acyclic resolutions. It follows easily that if $\mathrm{cd}\,\mathrm{Hom}(H^i(X), -) = t$ then an object in $\mathcal{D}(\mathcal{A})_{\leq -N}$ can be represented by an acyclic complex which is non-zero only in degree $\leq -N + t$. Hence it suffices to take $N > t$. □

REMARK 8.3.5. The reader may verify that the statement about $D(\mathcal{A})$ in the previous lemma is also true under the hypotheses AB4 and AB4*. This is more elegant, but less useful in practice.

COROLLARY 8.3.6. *Let $\mathcal{N}$ be the additive subcategory of* $\mathrm{Mod}(\tilde{X})$ *whose objects are direct sums of copies of $\mathcal{O}_L(-1)$. Then $\mathcal{N}$ is a thick subcategory (closed under extensions). Furthermore the map*

$$D(k) \xrightarrow{-\otimes \mathcal{O}_L(-1)} D_{\mathcal{N}}(\mathrm{Mod}(X))$$

is an equivalence.

PROOF. This follows from lemma 8.3.4 together with lemma 8.3.3. □

8.4. The main theorem

We prove the following Theorem 8.4.1. This is a non-commutative version of a theorem by Orlov [**27**]. Our proof is slightly different.

THEOREM 8.4.1. *There is a semi-orthogonal decomposition of* $D(\tilde{X})$ *given by* $(D(X), D(k))$. *The diagram corresponding to* (8.1) *is as follows*

$$D(k) \underset{F}{\overset{-\otimes_k \mathcal{O}_L(-1)}{\rightleftarrows}} D(\tilde{X}) \underset{L\alpha^*}{\overset{R\alpha_*}{\rightleftarrows}} D(X) \tag{8.10}$$

where F *is the left adjoint to* $-\otimes_k \mathcal{O}_L(-1)$ *(whose existence follows by lemma 8.1.1).*

In (8.10) *we may replace* D *by* D^* *where* $* = \emptyset, +, -, b$. *Furthermore we may also replace* D^* *everywhere by* D^*_f. *On* $D^-_f(\tilde{X})$, F *is given by* $\mathrm{RHom}(-, \mathcal{O}_L(-1))^*$.

PROOF. STEP 1. $R\alpha_* L\alpha^* = \mathrm{id}$. Since the objects in $\mathcal{V}$ are acyclic for α^*, this follows from Proposition 8.3.1.2 on $D^-(X)$.

The general case is then routine. If $\mathcal{F} \in D(X)$ then we have to show that the adjunction mapping $\mathcal{F} \to R\alpha_* L\alpha^* \mathcal{F}$ is a quasi-isomorphism, that is $H^i(\mathcal{F}) \to H^i(R\alpha_* L\alpha^* \mathcal{F})$ should be an isomorphism for all i. Since α^*, α_* have finite cohomological dimension, we can test this (for a fixed i) by replacing $\mathcal{F}$ by some $\tau_{\le N}\mathcal{F}$ for $N \gg 0$. But then $\mathcal{F} \in D^-(X)$ and this case was already covered.

STEP 2. The composition of $= - \otimes_k \mathcal{O}_L(-1)$ and $R\alpha_*$ is zero. This follows from the fact that $R\alpha_*$ commutes with direct sums (lemma 8.2.2) together with $R\alpha_* \mathcal{O}_L(-1) = 0$, which was proved in Proposition 8.3.1.

STEP 3. At this point $D(k)$ and $D(X)$ form a semi-orthogonal pair in $D(\tilde{X})$. To show that they form a semi-orthogonal decomposition we invoke lemma 8.1.4. Thus we have to prove that $\ker R\alpha_* = D(k)$. We prove first $\ker R\alpha_* = D_{\mathcal{N}}(\tilde{X})$ ($\mathcal{N}$ as in Cor. 8.3.6). Then by Corollary 8.3.6 we find $D_{\mathcal{N}}(\tilde{X}) = D(k)$.

We claim first that

$$\ker R\alpha_* \subset D_{\alpha^{-1}(\mathcal{C}_p)}(\tilde{X})$$

To prove this assume that $R\alpha_* \mathcal{M} = 0$. From the spectral sequence

$$R^p \alpha_* H^q(\mathcal{M}) \to R^{p+q} \alpha_* \mathcal{M}$$

we obtain short exact sequences

$$0 \to R^1 \alpha_* H^{i-1}(\mathcal{M}) \to R^i \alpha_* \mathcal{M} \to R^0 \alpha_* H^i(\mathcal{M}) \to 0$$

We conclude that for all i one has $\alpha_* H^i(\mathcal{M}) = 0$, whence by Proposition 6.5.2 $\mathcal{M} \in D_{\alpha^{-1}(\mathcal{C}_p)}(\tilde{X})$.

STEP 4. If $\phi : \mathcal{M} \to \mathcal{N}$ is a map in $D(\tilde{X})$ then we denote by $\mathrm{cone}(\phi)$ the cone of the triangle with base ϕ. $\mathrm{cone}(\phi)$ is unique up to non-unique isomorphism.

We claim

$$\ker R\alpha_* = \{\mathrm{cone}(L\alpha^* R\alpha_* \mathcal{M} \to \mathcal{M}) \mid \mathcal{M} \in D_{\alpha^{-1}(\mathcal{C}_p)}(\tilde{X})\} \tag{8.11}$$

Since $R\alpha_* L\alpha^* = \mathrm{id}$ it is clear that $RHS(8.11) \subset LHS(8.11)$. The opposite inclusion is obvious since if $R\alpha_* \mathcal{M} = 0$ then according to Step 3 : $\mathcal{M} \in D_{\alpha^{-1}(\mathcal{C}_p)}(\tilde{X})$. Hence $\mathcal{M} = \mathrm{cone}(L\alpha^* R\alpha_* \mathcal{M} \to \mathcal{M}) \in \mathrm{RHS}\,(8.11)$.

STEP 5. The formation of $\mathrm{cone}(L\alpha^* R\alpha_* \mathcal{M} \to \mathcal{M})$ is functorial and compatible with shifts and triangles.

According to lemma 8.1.4, we have a semi-orthogonal decomposition of $D(\tilde{X})$ given by $(\ker R\alpha_*, D(X))$. Since in the triangle

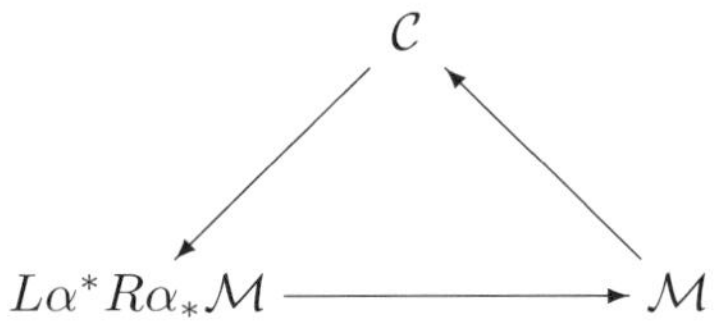

$L\alpha^* R\alpha_* \mathcal{M}$ lies in the essential image of $D(X)$ and $\mathcal{C} \in \ker R\alpha_*$ the good behaviour of $\operatorname{cone}(L\alpha^* R\alpha_* \mathcal{M} \to \mathcal{M})$ is a consequence of lemma 8.1.1.

STEP 6. According to Step 4 we have to show that if $\mathcal{M} \in D_{\alpha^{-1}(\mathcal{C}_p)}(\tilde{X})$ then $\operatorname{cone}(L\alpha^* R\alpha_* \mathcal{M} \to \mathcal{M}) \in D_{\mathcal{N}}(\tilde{X})$. That is

$$H^i(\operatorname{cone}(L\alpha^* R\alpha_* \mathcal{M} \to \mathcal{M})) \in \mathcal{N} \tag{8.12}$$

Now since $L\alpha^*$, $R\alpha_*$ have finite cohomological dimension the truth of (8.12) will not be influenced by the cohomology of $\mathcal{M}$ in high degree. Hence to verify (8.12) we may replace $\mathcal{M}$ by a suitable $\tau_{\leq a}\tau_{\geq b}\mathcal{M}$, that is, by a bounded complex. According to Step 5 and induction we may then assume that $\mathcal{M} \in \alpha^{-1}(\mathcal{C}_p)$.

Now (8.12) is a statement about homology of complexes, and it is easy to see that this homology is compatible with direct limits, when $\mathcal{M} \in \operatorname{Mod}(\tilde{X})$. Hence we may assume that $\mathcal{M} \in \alpha^{-1}(\mathcal{C}_p) \cap \operatorname{mod}(\tilde{X})$.

Now from lemma 8.4.2 below it will follow that we may in fact assume $\mathcal{M} = \mathcal{O}_L$, $\mathcal{O}_L(-1)$ or $\mathcal{O}_q$ with $q \in \beta^{-1}(O_\tau(p)) \setminus \beta^{-1}(\tau p)$. The last two cases are trivial since by Propositions 8.3.1 and 8.3.2 we find that $L\alpha^* R\alpha_* \mathcal{M} = \mathcal{M}$.

Hence assume $\mathcal{M} = \mathcal{O}_L$. Thus we have to compute

$$\operatorname{cone}(L\alpha^* R\alpha_* \mathcal{O}_L \to \mathcal{O}_L) = \operatorname{cone}(L\alpha^* \mathcal{O}_{\tau p} \to \mathcal{O}_L) \tag{8.13}$$

Since by Proposition 8.3.2 and $\alpha^* \mathcal{O}_{\tau p} = \mathcal{O}_L$, and α^* has cohomological dimension one we find that, up to shift, (8.13) is equal to $L_1\alpha^* \mathcal{O}_{\tau p}$, which according to Proposition 8.3.2 is equal to $\mathcal{O}_L(-1)$.

STEP 7. Odd and ends. First of all since $R\alpha_*$ and $L\alpha^*$ have finite cohomological dimension and preserve coherent objects (by Theorem 6.3.1), it is clear that the above reasoning may be repeated for D^* and for D^*_f with $* = \phi, +, -, b$.

To show that F on $D^-_f(\tilde{X})$ is equal to $\operatorname{RHom}(-, \mathcal{O}_L(-1))^*$ we have to show that this functor is well-defined and is a left adjoint to $- \otimes \mathcal{O}_L(-1)$. A quick verification shows that we have to check that $\operatorname{mod}(\tilde{X})$ contains enough objects $\mathcal{F}$ such that the homology of $\operatorname{RHom}_{o_X}(\mathcal{F}, \mathcal{O}_L(-1))$ is finite dimensional and concentrated in degree zero.

For $\mathcal{F}$ we take objects of the form $\alpha^*(\mathcal{E})(-n-1)$ with $n \in \mathbb{N}$ and $\mathcal{E} \in \mathcal{V} \cap \operatorname{mod}(X)$. It is easy to see that there are enough of those. We compute

$$\begin{aligned} \operatorname{RHom}_{o_{\tilde{X}}}(\alpha^*(\mathcal{E})(-n-1), \mathcal{O}_L(-1)) &= \operatorname{RHom}_{o_X}(\mathcal{E}, R\alpha_*(\mathcal{O}_L(n))) \\ &= \operatorname{RHom}_{o_X}(\mathcal{E}, (\mathcal{D}/m_{\tau p}\mathcal{D})_{n,o_X}) \end{aligned}$$

It now follows for example from Proposition 5.1.2 that the homology of the last line of the previous equation has the properties we want. $\square$

LEMMA 8.4.2. *$\alpha^{-1}(\mathcal{C}_p) \cap \operatorname{mod}(\tilde{X})$ is the smallest subcategory of* $\operatorname{Mod}(\tilde{X})$ *containing $\mathcal{O}_L(-1)$, $\mathcal{O}_L$ and $\mathcal{O}_q$ with $q \in \beta^{-1}(O_\tau(p)) \setminus \beta^{-1}(\tau p)$, and which is closed under*

1. *extensions;*
2. *kernels of epimorphisms;*
3. *cokernels of monomorphisms.*

PROOF. By Corollary 6.7.4 every object in $\alpha^{-1}(\mathcal{C}_p) \cap \mathrm{mod}(\tilde{X})$ is an extension of objects in $\mathrm{Mod}(L) \cap \mathrm{mod}(\tilde{X}) = \mathrm{mod}(L)$. Hence we have to verify the statement for $\mathrm{mod}(L)$. But then we may as well verify the corresponding statement for S. This is a routine exercise, which we leave to the reader. □

COROLLARY 8.4.3. *If* $\mathrm{Mod}(X)$ *has finite injective dimension then so does the blowup* $\mathrm{Mod}(\tilde{X})$.

PROOF. Assume that $\mathrm{Mod}(X)$ has finite injective dimension. We have to show that there exists an N such that $\mathrm{Hom}_{D(\tilde{X})}(\mathcal{M}, \mathcal{N}) = 0$ for all $\mathcal{M} \in D^b_{\geq N}(\tilde{X})$ and $\mathcal{N} \in D^b_{\leq 0}(\tilde{X})$.

For an arbitrary morphism $\phi : \mathcal{M} \to \mathcal{N}$ we have a commutative diagram of triangles

$$\begin{array}{ccccccc}
L\alpha^* R\alpha_* \mathcal{M} & \longrightarrow & \mathcal{M} & \longrightarrow & F\mathcal{M} \otimes_k \mathcal{O}_L(-1) & \longrightarrow \\
{\scriptstyle L\alpha^* R\alpha_*(\phi)} \downarrow & & {\scriptstyle \phi} \downarrow & & {\scriptstyle F\phi \otimes_k \mathcal{O}_L(-1)} \downarrow & \\
L\alpha^* R\alpha_* \mathcal{N} & \longrightarrow & \mathcal{N} & \longrightarrow & F\mathcal{N} \otimes_k \mathcal{O}_L(-1) & \longrightarrow
\end{array}$$

Since $L\alpha^*$ is fully faithful and $\mathrm{Mod}(X)$ has finite injective dimension we find that $L\alpha^* R\alpha_*(\phi) = 0$ if N is large enough. Similarly $D^b(k)$ is a semi-simple category and hence $F\phi = 0$ if N is large enough. This then implies that ϕ is in fact obtained from a map $\theta : F\mathcal{M} \otimes_k \mathcal{O}_L(-1) \to L\alpha^* R\alpha_* \mathcal{N}$. However by lemma 8.3.3 we find that $\mathrm{Hom}_{D(\tilde{X})}(\mathcal{O}_L, -)$ has finite cohomological dimension. Hence by making N even larger if necessary we also find that $\theta = 0$. This finishes the proof. □

CHAPTER 9

Some results on graded algebras and their sections

9.1. Generalities

In this section $\gamma : X \to \operatorname{Spec} k$ will be a noetherian quasi-scheme over k which is proper (see Def. 3.8.4). As usual we put $\mathcal{O}_X = \gamma^* k$. We will also assume that $\mathcal{O}_X$ is noetherian. By adjointness we have $\gamma_* \mathcal{M} = \operatorname{Hom}_{\operatorname{Mod}(X)}(\mathcal{O}_X, \mathcal{M}) = \Gamma(X, \mathcal{M})$. Similarly for the derived functors. Hence the properness of X simply means that if $\mathcal{M} \in \operatorname{mod}(X)$ then $H^i(X, \mathcal{M})$ is a finite dimensional k-vector space for all i. The fact that $\mathcal{O}_X$ is noetherian implies that $H^i(X, -)$ commutes with direct limits.

Assume that $\mathcal{E}$ is a noetherian $\mathbb{N}$-graded algebra on X and let $E = \Gamma(X, \mathcal{E})$ (cfr §3.5). Then $(\mathcal{E})_{o_X}$ is an $E - \mathcal{E}$-bimodule, in the obvious sense. We have a "global section functor"

$$\Gamma(X, -) : \operatorname{Gr}(\mathcal{E}) \to \operatorname{Gr}(E) \tag{9.1}$$

and we denote its left adjoint by

$$- \otimes_E \mathcal{E}_{o_X} : \operatorname{Gr}(E) \to \operatorname{Gr}(\mathcal{E}) \tag{9.2}$$

As the notation suggests $- \otimes_E \mathcal{E}_{o_X}$ is the functor commuting with shifts and colimits which sends E to $\mathcal{E}_{o_X}$.

Our aim will be to relate $\operatorname{QGr}(\mathcal{E})$ to $\operatorname{QGr}(E)$. To this end we introduce the following definition.

DEFINITION 9.1.1. $\mathcal{E}$ is ample if for every object $\mathcal{M}$ in $\operatorname{mod}(X)$ we have for $n \gg 0$ that $\mathcal{M} \otimes_{o_X} \mathcal{E}_n$ is generated by global sections and $H^i(X, \mathcal{M} \otimes_{o_X} \mathcal{E}_n) = 0$ for $i > 0$.

The following proposition was essentially proved in [**10, 34**] (see [**34**, Thm 5.2]) under more restrictive hypotheses.

PROPOSITION 9.1.2. *Let $\mathcal{E}$ be as above and assume that $\Gamma(X, -)$ has finite cohomological dimension. If $\mathcal{E}$ is ample then E is noetherian. The functors in* (9.1) *and* (9.2) *send* $\operatorname{gr}(E)$ *to* $\operatorname{gr}(\mathcal{E})$ *and vice-versa. Furthermore these functors define inverse equivalences between* $\operatorname{QGr}(E)$ *and* $\operatorname{QGr}(\mathcal{E})$.

Unfortunately ampleness of $\mathcal{E}$ is too strong for the applications we have in mind. However we will show that under some extra conditions the functors (9.1) and (9.2) may still be well behaved, even if $\mathcal{E}$ is non-ample.

We introduce the following hypotheses.

HYPOTHESIS (***). $\Gamma(X, -)$ *has finite cohomological dimension and furthermore there is an injective graded $\mathcal{E}$-bimodule map $t : \mathcal{E}(-1) \to \mathcal{E}$ such that $\mathcal{E}/\operatorname{im} t$ is ample. Furthermore the induced map $\mathcal{E}_{o_X}(-1) \to \mathcal{E}_{o_X}$ is monic.*

Taking global sections yields a regular central element of E in degree one which we also denote by t. We write $\bar{\mathcal{E}} = \mathcal{E}/\operatorname{im} t$. If $\mathcal{N} \in \operatorname{Gr}(\mathcal{E})$ then t defines a map

$\mathcal{N}(-1) \to \mathcal{N}$ in $\mathrm{Gr}(\mathcal{E})$. We put $\bar{\mathcal{N}} = \mathcal{N}/\operatorname{im} t$. We say that $\mathcal{N}$ is annihilated by t if $t : \mathcal{N}(-1) \to \mathcal{N}$ is the zero map. We say that $\mathcal{N}$ is t-torsion if $\mathcal{N}$ is a union of objects which are each annihilated by some t^n (n variable). We say that $\mathcal{N}$ is t-torsion free if t is monic. Similar conventions apply to $\mathrm{QGr}(\mathcal{E})$, $\mathrm{Gr}(E)$ and $\mathrm{Qgr}(E)$.

PROPOSITION 9.1.3. *Let $\mathcal{N} \in \mathrm{gr}(\mathcal{E})$. Then*

1. *Multiplication by t on $H^i(X, \mathcal{N})$ is an automorphism in high degree when $i > 0$.*
2. *The complex*
$$\Gamma(X, \mathcal{N}(-1)) \xrightarrow{t} \Gamma(X, \mathcal{N}) \to \Gamma(X, \bar{\mathcal{N}}) \to 0$$
is exact in high degree.
3. *$H^i(X, \mathcal{N})$ is finitely generated for all i.*

Furthermore we also have

4. *$\mathrm{Tor}_i^E(-, \mathcal{E})$ sends $\mathrm{Tors}(E)$ to $\mathrm{Tors}(\mathcal{E})$ for all i.*
5. *E is noetherian.*

PROOF. 1.,2. Using the appropriate long exact sequences it follows that it is sufficient to prove this in the case that $\mathcal{N}$ is annihilated by t and in the case that $\mathcal{N}$ is t-torsion free.

In the first case we have that $\mathcal{N} = \bar{\mathcal{N}}$ and furthermore, for $i > 0$, $H^i(X, \bar{\mathcal{N}})$ is right bounded (by ampleness of $\bar{\mathcal{E}}$). Hence 1.,2. are trivially true in this case.

Let us consider the second case. We apply the long exact sequence for $\Gamma(X, -)$ to
$$0 \to \mathcal{N}(-1) \to \mathcal{N} \to \bar{\mathcal{N}} \to 0$$
Since $H^i(X, \bar{\mathcal{N}})$ is right bounded, we immediately obtain the statement about $H^i(X, \mathcal{N})$ for $i > 1$. So let us assume $i \leq 1$. For $n \gg 0$ we have an exact sequence
$$0 \to \Gamma(X, \mathcal{N}_{n-1}) \xrightarrow{t} \Gamma(X, \mathcal{N}_n) \to \Gamma(\bar{\mathcal{N}}_n) \to H^1(X, \mathcal{N}_{n-1}) \xrightarrow{t} H^1(X, \mathcal{N}_n) \to 0$$
So $\dim H^1(X, \mathcal{N}_n)$ is descending and hence must eventually become constant. Thus $t : H^1(X, \mathcal{N}_{n-1}) \to H^1(X, \mathcal{N}_n)$ is an isomorphism for $n \gg 0$. This proves 1.,2. for $i = 0, 1$.

3. In view of 1., the only non-trivial case is $i = 0$. By Proposition 9.1.2 $\Gamma(X, \bar{\mathcal{N}})$ is finitely generated. Since according to 2., $\overline{\Gamma(X, \mathcal{N})}$ has finite colength in $\Gamma(X, \bar{\mathcal{N}})$ we find that $\overline{\Gamma(X, \mathcal{N})}$ is finitely generated. Hence by the graded version of Nakayama's lemma, $\Gamma(X, \mathcal{N})$ is also finitely generated.
4. Since Tor is compatible with direct limits we may prove the corresponding statement for "tors". Furthermore, using the appropriate long exact sequences, it follows that it is sufficient to prove that $\mathrm{Tor}_i^E(V, \mathcal{E})$ is right bounded for a finite dimensional E_0-module V (considered as E-module).

Since V is annihilated by t, we have as usual
$$\mathrm{Tor}_i^E(V, \mathcal{E}) = \mathrm{Tor}_i^{\bar{E}}(V, \bar{\mathcal{E}}) \tag{9.3}$$
Put $\bar{E}' = \Gamma(X, \bar{\mathcal{E}})$. According to 2., applied to $\mathcal{E}_{o_X}$ we find that the map $\bar{E}' \to \bar{E}$ is monic and has finite dimensional cokernel. Hence by [**6**, Prop. 2.5] and Proposition 9.1.2 one has $\mathrm{QGr}(\bar{E}') = \mathrm{QGr}(\bar{E}) = \mathrm{QGr}(\bar{\mathcal{E}})$. The

functor realizing this equivalence is given by $-\otimes_{\bar{E}'}\bar{\mathcal{E}}$. Hence this is in particular an exact functor. Now to compute the righthand side of (9.3) we take a free resolution of V. Since the grading on V is right bounded this free resolution is an exact sequence in $\operatorname{QGr}(\bar{E}')$. Hence it remains exact after applying $-\otimes_{\bar{E}'}\bar{\mathcal{E}}$. In this way we obtain that $\operatorname{Tor}_i^E(V,\mathcal{E})\in\operatorname{Tors}(\mathcal{E})$ for all i.

5. By Proposition 9.1.2 it follows that $\bar{E}'$ is noetherian. Since the map $\bar{E}\to\bar{E}'$ is injective and has finite dimensional cokernel, we deduce that $\bar{E}$ is also noetherian. Whence, by a Hilbert-type argument, E is noetherian. □

For use below we write $W=\operatorname{Proj}\mathcal{E}$, $T=\operatorname{Proj}\bar{\mathcal{E}}$, $V=\operatorname{Proj}E$, $S=\operatorname{Proj}\bar{E}$. It follows that S,T are S divisors in W,V.

As usual, we will denote the quotient functors $\operatorname{Gr}(E)\to\operatorname{QGr}(E)$, $\operatorname{Gr}(\mathcal{E})\to\operatorname{QGr}(\mathcal{E})$ by π.

PROPOSITION 9.1.4. *The functors*

$$\delta^*:\operatorname{Mod}(V)\to\operatorname{Mod}(W):\pi M\mapsto\pi(M\otimes_E\mathcal{E})$$
$$\delta_*:\operatorname{Mod}(W)\to\operatorname{Mod}(V):\pi N\mapsto\pi\Gamma(X,N)$$

are well defined and form an adjoint pair.

In particular (δ^*,δ_*) *defines an morphism of quasi-schemes*

$$\delta:W\to V$$

δ *fits in a commutative diagram of quasi-schemes*

$$\begin{array}{ccc} T & \xrightarrow{\ j\ } & W \\ {\scriptstyle\delta}\downarrow & & \downarrow{\scriptstyle\delta} \\ S & \xrightarrow{\ i\ } & V \end{array} \tag{9.4}$$

Here i,j *are the inclusion mappings.* $\delta:T\to S$ *is an isomorphism. If* $\mathcal{M}\in\operatorname{Mod}(T)$ *then we have*

$$\delta_*(\mathcal{M}\otimes_{o_T}\mathcal{N}_{T/W})=\delta_*(\mathcal{M})\otimes_{o_S}\mathcal{N}_{S/V} \tag{9.5}$$

Thus the normal bundles on T *and* S *correspond to each other under the isomorphism* δ.

PROOF. It is obvious that δ_* is well-defined. The fact that δ^* is well-defined follows easily from Proposition 9.1.3.4.

We now show that δ_*, δ^* are adjoint functors. This means that we have to construct a natural isomorphism between

$$\operatorname{Hom}_{\operatorname{QGr}(E)}(\mathcal{M},\delta_*\mathcal{N})$$

and

$$\operatorname{Hom}_{\operatorname{QGr}(\mathcal{E})}(\delta^*\mathcal{M},\mathcal{N})$$

where $\mathcal{M}=\pi M$, $\mathcal{N}=\pi N$.

By taking a presentation of M we reduce to $M=E$. We have

$$\begin{aligned}\operatorname{Hom}_{\operatorname{QGr}(E)}(E,\delta_*\mathcal{N})&=\varinjlim_n\operatorname{Hom}_{\operatorname{Gr}(E)}(E_{\geq n},\Gamma(X,N))\\&=\varinjlim_n\operatorname{Hom}_{\operatorname{Gr}(\mathcal{E})}(E_{\geq n}\otimes_E\mathcal{E}_{o_X},N)\end{aligned}$$

and we have to show that this is equal to

$$\varinjlim_n \operatorname{Hom}_{\operatorname{Gr}(\mathcal{E})}((\mathcal{E}_{oX})_{\geq n}, N)$$

Put

$$K_n = \ker(E_{\geq n} \otimes_E \mathcal{E}_{oX} \to (\mathcal{E}_{oX})_{\geq n})$$
$$C_n = \operatorname{coker}(E_{\geq n} \otimes_E \mathcal{E}_{oX} \to (\mathcal{E}_{oX})_{\geq n})$$

It is now sufficient to show that K_n, C_n are *torsion* inverse systems. That is for all n there exists an m such that the maps $K_{n+m} \to K_n$, $C_{n+m} \to C_n$ are zero.

Tensoring the exact sequence

$$0 \to E_{\geq n} \to E \to E/E_{\geq n} \to 0$$

with $\mathcal{E}_{oX}$ and restricting to degrees $\geq n$ yields that

$$K_n = \operatorname{Tor}_1^E(E/E_{\geq n}, \mathcal{E}_{oX})$$
$$C_n = (E/E_{\geq n} \otimes_E \mathcal{E}_{oX})_{\geq n}$$

In particular it follows from Proposition 9.1.3.4 that K_n, C_n are right bounded. Furthermore it is clear that C_n is zero in degrees $< n$, and by considering a minimal free resolution of $E/E_{\geq n}$ we see that the same holds for K_n. It now follows clearly that K_n, C_n are torsion inverse systems.

That (9.4) is a commutative diagram is obvious and the fact that $\delta : T \to S$ is an isomorphism is simply the ampleness of $\bar{\mathcal{E}}$.

The identity (9.5) is just a special case of the fact that δ_* commutes with shift. □

From Proposition 9.1.3 one easily deduces that the shifts of πE are acyclic for δ^* (in the sense of [**18**]). There are clearly enough of those, so $L\delta^*$ exists.

It is easy to verify the following formulas

$$R^i\delta_*\pi N = \pi H^i(X, N) \tag{9.6}$$
$$L_i\delta^*\pi M = \pi \operatorname{Tor}_i^E(M, \mathcal{E})$$

LEMMA 9.1.5. *$R\delta_*$ sends $D_f^+(W)$ to $D_f^+(V)$ and $L\delta^*$ sends $D_f^-(V)$ to $D_f^-(W)$.*

PROOF. The statement about $L\delta^*$ follows from the fact that E and $\mathcal{E}$ are noetherian. The statement about $R\delta_*$ follows from Proposition 9.1.3.3. □

Thus in particular δ is *proper* in the sense of Definition 3.8.4.

The following is also easy.

PROPOSITION 9.1.6. *If $\mathcal{N} \in \operatorname{Tors}_T(W)$ then $R^i\delta_*\mathcal{N} = 0$ for $i > 0$. Similary if $\mathcal{M} \in \operatorname{Tors}_S(V)$ then $L^i\delta^*\mathcal{M} = 0$ for $i > 0$. Finally δ_*, δ^* define inverse equivalences between $\operatorname{Tors}_T(W)$ and $\operatorname{Tors}_S(V)$ (see §3.6 for notation).*

PROOF. Since all functors involved commute with direct limits we may assume that $\mathcal{M}$, $\mathcal{N}$ are coherent. The first two statements are easily seen to be true for objects annihilated by t. The general case follows from this by considering the appropriate long exact sequences.

In particular we obtain that δ_* and δ^* are exact. We now have to show that the adjunction morphisms are isomorphisms on coherent objects. Again this is true for objects annihilated by t by the hypotheses on $\bar{\mathcal{E}}$. The general case now follows by filtering objects in such a way that the associated graded quotients are annihilated by t. $\square$

The following proposition will be true in all our applications for trivial reasons. However it may be interesting to note that one can prove it in this generality.

PROPOSITION 9.1.7. $\mathrm{Mod}(W)$ *has enough T-torsion free objects. In particular Lj^* is defined.*

PROOF. It suffices to show that every object in $\mathrm{mod}(W)$ is a quotient of a T-torsion free one. Let $\mathcal{M} = \pi M$ where $M \in \mathrm{gr}\,\mathcal{E}$. With an argument as in the proof of lemma 3.6.4 one finds a $M_0 \subset M$ which is t-torsion free such that M/M_0 is t-torsion. Assume $t^n(M/M_0) = 0$ and consider the following diagram with exact rows.

$$\begin{array}{ccccccccc} 0 & \longrightarrow & M_0 & \longrightarrow & M & \longrightarrow & M/M_0 & \longrightarrow & 0 \\ & & \downarrow & & \downarrow & & \| & & \\ 0 & \longrightarrow & M_0/t^nM_0 & \longrightarrow & M/t^nM & \longrightarrow & M/M_0 & \longrightarrow & 0 \end{array} \tag{9.7}$$

According to Propositin 9.1.3,

$$\Gamma(X, M) \to \Gamma(X, M/t^nM) \tag{9.8}$$

is epic in high degree. Hence there exists an epimorphism $E(-a)^b \to \Gamma(X, M)$ for certain a, b such that the composition with (9.8) has finite cokernel.

Tensoring with $\mathcal{E}_{o_X}$ and composing with $\Gamma(X, M) \otimes_E \mathcal{E}_{o_X} \to M$ we obtain a map $\mathcal{E}_{o_X}(-a)^b \to M$ such that the composition with $M \to M/t^nM$ has right bounded cokernel (using Propositions 9.1.3.4 and 9.1.6). Applying π to the induced map $M_0 \oplus \mathcal{E}_{o_X}(-a)^b \to M$ together with (9.7) yields what we want. $\square$

PROPOSITION 9.1.8. *All morphisms and compositions in* (9.4) *are admissible. Furthermore one has the following identities.*

$$i^*\delta_* = \delta_* j^* \tag{9.9}$$

$$i^!\delta_* = \delta_* j^! \tag{9.10}$$

$$Ri^!\, R\delta_* = R\delta_* Rj^! \tag{9.11}$$

$$Li^*\, R\delta_* = R\delta_* Lj^* \tag{9.12}$$

PROOF. Checking admissibility is routine using the explicit formulas (9.6). We leave this to the reader.

To verify the compatibility $i^*\delta_* = \delta_* j^*$ we may work in $\mathrm{mod}(X)$ (since everything is compatible with direct limits). But then it is just a reformulation of Proposition 9.1.3.2.

The other compatibility $i^!\delta_* = \delta_* j^!$ involves only left exact functors, so it may be verified on injectives. Let $F \in \mathrm{Mod}(W)$ be such an injective. We have an exact sequence

$$0 \to j^!F \to F \to F(T) \to 0$$

Now $j^!F \in \mathrm{Tors}_T(W)$ hence $R^1\delta_* j^!F = 0$. Therefore we obtain an exact sequence

$$0 \to \delta_* j^!F \to \delta_* F \to \delta_*(F(T)) \to 0 \tag{9.13}$$

Now δ_* is compatible with shift. That is $\delta_*(F(T)) = (\delta_* F)(S)$. Since we also have an exact sequence

$$0 \to i^! \delta_* F \to \delta_* F \to (\delta_* F)(S) \tag{9.14}$$

we are through.

(9.11) follows from (9.10) provided we show for every injective $E \in \operatorname{Mod}(X)$ that $j^! E$ is acyclic for δ_* and that $\delta_* E$ is acyclic for $i^!$.

The statement about $j^! E$ is clear since $j^! E$ is supported on T. The statement of $\delta_* E$ follows from the right exactness of (9.14). However (9.14) is the same as (9.13) and the latter is right exact. This shows what we want.

(9.12) follows from (9.11) using (7.2). □

In the next few paragraphs we investigate to what extent the morphism $\delta : W \to V$ is an isomorphism.

THEOREM 9.1.9. 1. *$\operatorname{Iso}_S(V)$ and $\operatorname{Iso}_T(W)$ are mapped to each other under (δ^*, δ_*).*
2. *$R^i \delta_*$ sends $\operatorname{Mod}(W)$ to $\operatorname{Iso}_S(V)$ for $i > 0$.*
3. *$L_i \delta^*$ sends $\operatorname{Mod}(V)$ to $\operatorname{Iso}_T(W)$ for $i > 0$.*
4. *(δ^*, δ_*) define inverse equivalences between the categories $\operatorname{Mod}(W)/\operatorname{Iso}_T(W)$ and $\operatorname{Mod}(V)/\operatorname{Iso}_S(V)$.*

PROOF. 1. This is clear by functoriality.
2. This is precisely 1. of Proposition 9.1.3.
3. We may assume that $\mathcal{M}$ is coherent. Furthermore by considering the appropriate long exact sequence it suffices to consider the cases where $\mathcal{M}$ is t-torsion and where $\mathcal{M}$ is t-torsion free.

   If $\mathcal{M}$ is t-torsion then $L_i \delta^* \mathcal{M} = 0$ for $i > 0$ by Proposition 9.1.6 so the result is true. Hence assume that $\mathcal{M}$ is t-torsion free. Then the long exact sequence for δ^* applied to

$$0 \to \mathcal{M}(-1) \to \mathcal{M} \to \bar{\mathcal{M}} \to 0$$

   yields that t is an isomorphism between $L_i \delta^* \mathcal{M}(-1)$ and $L_i \delta^* \mathcal{M}$ for $i > 0$, proving 3. in this case also.
4. We have to show that the adjunction mappings are isomorphisms. We already know that δ_*, δ^* are exact functors between $\operatorname{Mod}(W)/\operatorname{Iso}_T(W)$ and $\operatorname{Mod}(V)/\operatorname{Iso}_S(V)$ commuting with direct limits. Hence it suffices to show that the adjunction mappings are isomorphisms for objects annihilated by t and for objects which are t-torsion free.

   The case of objects annihilated by t is already covered by Proposition 9.1.6, so we concentrate on t-torsion free objects.

   First assume that $\mathcal{N} \in \operatorname{mod}(W)$ is a t-torsion free object. We have an exact sequence

$$0 \to \mathcal{N}(-1) \to \mathcal{N} \to \bar{\mathcal{N}} \to 0 \tag{9.15}$$

   Working modulo $\operatorname{iso}_W(T)$ this yields an exact sequence

$$0 \to \delta^* \delta_* \mathcal{N}(-1) \to \delta^* \delta_* \mathcal{N} \to \delta^* \delta_* \bar{\mathcal{N}} \to 0 \tag{9.16}$$

   Now by Proposition 9.1.6 we already know that $\delta^* \delta_* \bar{\mathcal{N}} = \bar{\mathcal{N}}$. Combining (9.15) and (9.16) using the snake lemma now yields what we want.

The proof that the other adjunction mapping is an isomorphism is entirely similar. □

9.2. The case of a blowing up

Now we let X, Y, p be as in §5. Furthermore we also assume that X/k is proper.

Up to now we have developed everything without assuming projectivity. However in this section we throw in the towel and we introduce the following hypothesis.

HYPOTHESIS (****). *p is a smooth point on Y. $o_X(Y)$ is ample on X and the invertible o_Y-bimodule I_Y is ample on Y.*

The fact that X has an ample invertible bimodule means in particular that Hypothesis (**) is satisfied. Furthermore it is also easy to check that under the hypotheses one has $\operatorname{cd}\Gamma(X,-) \leq 2$. This is part of Hypothesis (***).

The smoothness of Y in p implies that $\tilde{Y} \to Y$ is an isomorphism (see Theorem 6.3.1). Also because of smoothness, we have by (6.16) a monomorphism $t : \mathcal{D}(-1) \to \mathcal{D}$ such that $\bar{\mathcal{D}} = \mathcal{D}/\operatorname{im} t$ is equal to $\mathcal{D}_Y$ in degree ≥ 1. By lemma 8.2.1 it is also clear that t remains monic after applying the functor $(-)_{o_X}$.

Since I_Y is ample on Y we find that $\mathcal{D}_Y$ and hence $\bar{\mathcal{D}}$ is ample. Thus Hypothesis (***) holds and hence the material in §9.1 applies to the current situation. Let D, D_Y be the global sections of $\mathcal{D}$ and $\mathcal{D}_Y$ respectively and put $V = \operatorname{Proj} D$. We now have a commutative diagram of k-quasi-schemes

$$\begin{array}{ccccc} & & Y & & \\ & \overset{i}{\swarrow} & \downarrow{\scriptstyle i} & \overset{i}{\searrow} & \\ X & \xleftarrow[\alpha]{} & \tilde{X} & \xrightarrow[\delta]{} & V \end{array} \tag{9.17}$$

where the arrows marked i are divisors. All morphisms and compositions of maps in this diagram are admissible by lemma 7.2.12, Proposition 9.1.8 and lemma 7.2.4. Further properties of δ maybe deduced from §9.1. We define $\mathcal{O}_M = \delta_* \mathcal{O}_L$ and we will refer to $\mathcal{O}_M$ as the exceptional curve in V.

Since we are in a specific situation here, we can of course prove stronger results than those in §9.1. First of all note that we have the following.

PROPOSITION 9.2.1. *D satisfies χ* [**6**, Def. 3.7].

PROOF. Denote the global sections of $\mathcal{D}_Y$ by D_Y. Since I_Y is ample, D_Y satisfies χ [**6**]. Furthermore, by Proposition 9.1.3.2 the monomorphism $\bar{D} \to D_Y$ has finite cokernel. Hence by [**6**, Lemma 8.2], $\bar{D}$ will also satisfy χ. Then we can apply [**6**, Thm 8.8] to obtain that D also satisfies χ. □

We also have the following important result.

PROPOSITION 9.2.2. *The cohomological dimension of δ_* is less than or equal to one. Furthermore if $\mathcal{N} \in \operatorname{gr}(\mathcal{D})$ then $R^1\delta_*\mathcal{N}$ is a finite extension of quotients of $\mathcal{O}_M$.*

PROOF. Let $\mathcal{N} \in \operatorname{gr}(\mathcal{D})$. By taking a resolution of $\mathcal{N}$ we see that it is sufficient to treat the case $\mathcal{N} = \mathcal{M} \otimes_{o_X} \mathcal{D}$ where $\mathcal{M} \in \operatorname{mod}(X)$. Now define $\mathcal{E} = \oplus_{n \geq 0} o_X(nY)$. It is easy to see that $\mathcal{D}$ embeds in $\mathcal{E}$ as graded $\mathcal{D}$-bimodule. Put $\mathcal{N}' = \mathcal{M} \otimes_{o_X} \mathcal{E}$. Let

K, I, C be respectively the kernel, image and cokernel of the induced map $\mathcal{N} \to \mathcal{N}'$. Thus we have exact sequences

$$0 \to K \to \mathcal{N} \to I \to 0$$
$$0 \to I \to \mathcal{N}' \to C \to 0$$

It immediately follows from Theorem 5.5.10 that K, C are in $\mathcal{C}_p(\mathcal{D})$. Hence from (9.6) it follows that $R^i\delta_*\pi K = R^i\delta_*\pi C = 0$ for $i > 0$. Since $o_X(Y)$ is ample it also follows from (9.6) that $R^i\delta_*\pi\mathcal{N}' = 0$ for $i > 0$. Plugging this in the above exact sequences we find that $R^i\delta_*\pi\mathcal{N} = 0$ for $i > 1$ and also that $R^1\delta_*\pi\mathcal{N}$ is a quotient of $\delta_*\pi C$.

Now C itself is not necessarily coherent, but it is a union of C_i which are coherent. Since δ_* is compatible with direct limits we deduce that $R^1\delta_*\pi\mathcal{N}$ is a quotient of some $\delta_*\pi C_i$. From lemma 6.7.3 it follows that πC_i has a finite filtration whose associated graded quotients are quotients of $\mathcal{O}_L(t)$. It then follows that $\delta_* C_i$ has a finite filtration whose associated graded quotients are quotients of $\mathcal{O}_M(t)$. Since these quotients are in $\operatorname{iso}_Y(V)$ they are invariant under shifting. Hence they are also quotients of $\mathcal{O}_M$. The result now follows for $R^1\delta_*\pi\mathcal{N}$. □

The hypotheses for the following proposition will hold in our applications.

PROPOSITION 9.2.3. *Assume that in addition to Hypotheses (****) we have* $R\Gamma(X, \mathcal{O}_X) = k$, $H^1(Y, I^n_{Y,p}) = 0$ *for all* n *and* $\Gamma(Y, \mathcal{O}_Y) = k$. *Then*

1. $R\Gamma(X, \mathcal{D}_{o_X}) = D$.
2. *There is an exact sequence*
$$0 \to D(-1) \xrightarrow{t} D \to D_Y \to 0$$
3. $R\delta_* L\delta^* = \mathrm{id}$ *on* $D^-(V)$.
4. $R\Gamma(\tilde{X}, \mathcal{O}_{\tilde{X}}) = k$
5. $R\Gamma(V, \mathcal{O}_V) = k$.

PROOF. 1.,2. We have $\mathcal{D}_0 = \mathcal{O}_X$ and furthermore by (6.16) there are exact sequences for $n \geq 1$
$$0 \to \mathcal{D}_{n-1} \to \mathcal{D}_n \to I^n_{Y,p} \to 0$$
Looking at the corresponding long exact sequence for $\Gamma(X, -)$ yields what we want by induction.
3. This follows from 1. since $L\delta^*$ can be computed by free resolutions. Note that $R\delta_*$ is defined on $D^-(\tilde{X})$ since according to Proposition 9.2.2, δ_* has finite cohomological dimension.
4. We have
$$\begin{aligned} R\Gamma(\tilde{X}, \mathcal{O}_{\tilde{X}}) &= R\Gamma(X, R\alpha_*\mathcal{O}_{\tilde{X}}) \qquad \text{(admissibility)} \\ &= R\Gamma(X, \mathcal{O}_X) \qquad \text{(Proposition 8.3.1.2).} \\ &= k \qquad \text{(by hypothesis)} \end{aligned}$$
5. We have
$$\begin{aligned} R\Gamma(V, \mathcal{O}_V) &= R\Gamma(V, R\delta_*\mathcal{O}_{\tilde{X}}) \qquad \text{(by 1.)} \\ &= R\Gamma(\tilde{X}, \mathcal{O}_{\tilde{X}}) \qquad \text{(admissibility)} \\ &= k \qquad \text{(by 4.)} \quad \square \end{aligned}$$

CHAPTER 10

Quantum plane geometry

In this section we fix a "triple" $(Y, \sigma, \mathcal{L})$ with Y a smooth elliptic curve of degree three in $\mathbb{P}^2$, $\sigma \in \operatorname{Aut}(Y)$ a translation and $\mathcal{L} = i^*\mathcal{O}_{\mathbb{P}^2}(1)$ where $i : Y \to \mathbb{P}^2$ denotes the inclusion.

We let $A = A(Y, \sigma, \mathcal{L})$ be the regular algebra associated to this triple [**8**] and we put $X = \operatorname{Proj} A$. Since A has the Hilbert series of a three dimensional polynomial algebra it is customary to view X as a non-commutative $\mathbb{P}^2$.

A contains a central element g in degree three (determined up to a scalar) such that $Y = \operatorname{Proj} A/gA$ [**8, 9, 10**]. Clearly $Y \subset X$ is a divisor and so the theory of the previous sections applies to it.

Since $X = \operatorname{Proj} A$, $\operatorname{Mod}(X)$ carries a canonical shift functor which we denote by $\mathcal{M} \mapsto \mathcal{M}(1)$. The corresponding invertible bimodule is denoted by $o_X(1)$. Thus $o_X(Y) = o_X(1)^{\otimes 3} \overset{\text{not.}}{=} o_X(3)$. The o_Y bimodule $o_Y(1)$ is defined similarly. By construction we have $o_Y(1) = \mathcal{L}_\sigma$.

In addition we use the functors ω, π, $\tilde{-}$ which were defined in §3.7.

10.1. Multiplicities of some objects

If $\mathcal{F} \in \operatorname{mod}(X)$ then $\Gamma(X, \mathcal{F})$ is finite dimensional. We denote by $\operatorname{Ann}_{\mathcal{O}_X} \mathcal{F}$ the kernel of the map $\mathcal{O}_X \to \Gamma(X, \mathcal{F})^* \otimes_k \mathcal{F}$ which is obtained from the map $\Gamma(X, \mathcal{F}) \otimes_k \mathcal{O}_X \to \mathcal{F}$. By using the exactness properties of the completion functor (§5) it follows easily that if $\mathcal{F} \in \mathcal{C}_f$ (cfr §5) then we have

$$\text{(10.1)} \qquad \operatorname{length}(\mathcal{O}_X / \operatorname{Ann}_{\mathcal{O}_X} \mathcal{F}) = \oplus_{p \in Y/\langle\tau\rangle} \dim_k (C_p / \operatorname{Ann}_{C_p}(\hat{\mathcal{F}}_p))$$

Let $\mathcal{F} \in \operatorname{trans}_Y(X)$ and let $p \in Y$. Then we define numbers $t_{p,n}$ for $n \geq 1$ by $T_p(\mathcal{F}) = \oplus_{n \geq 1} \mathcal{O}_{Y,p}/m_{Y,p}^{t_{p,n}}$ (note that to simplify the notation we have dropped the completion sign). We consider $(t_{p,n})_n$ as a partition of the length of $T_p(\mathcal{F})$ and we let $(r_{p,n}(\mathcal{M}))_n$ stand for the conjugate partition. In the sequel we will consider the (p, n) as the points which are infinitely near to p. The numbers $r_{p,n}(\mathcal{M})$ should be viewed as the multiplicities of those infinitely near points. We identify $(p, 1)$ with p.

If $\mathcal{M} \in \operatorname{mod}(X)$ then $\mathcal{M} = \pi M$ for some finitely generated A-module M. As usual [**9**] we have $\dim M_n = (e/d!)n^d + f(n)$ for $n \gg 0$ where d and e are respectively the Gelfand-Kirillov dimension and the multiplicity of M and where $f(n)$ is a polynomial of degree $< d$. We write $d(\mathcal{M}) = d - 1$ and $e(\mathcal{M}) = e$.

We conjecture the following (cfr [**19**, Cor V.3.7, p389]).

CONJECTURE 10.1.1. Let $\mathcal{M} \in \operatorname{trans}_Y(X)$ and assume that the image of $\mathcal{M}$ in $\operatorname{mod}(X)/\mathcal{C}_f$ is simple. Write $e = e(\mathcal{M})$ and $r_{p,n} = r_{p,n}(\mathcal{M})$ for $p \in Y$ and $n \geq 1$.

Then the following inequality holds:

$$\sum_{p,n} \frac{r_{p,n}(r_{p,n}-1)}{2} \leq \frac{(e-1)(e-2)}{2} \tag{10.2}$$

Note that the modern proof of the commutative analogue of this conjecture uses resolution of singularities for a curve through blowing up the ambient projective plane, together with the fact that an irreducible curve has non-negative genus. Unfortunately we have not been able to generalize this proof to the non-commutative case. Indeed it is rather the other way round. We would like to use a result as (10.2) to deduce properties of our non-commutative blowing up.

Imitating one of the classical proofs of (10.2) (see [**15**]) as far as possible leads to Proposition 10.1.2 below.

PROPOSITION 10.1.2. *Let $\mathcal{M} \in \operatorname{trans}_Y(X)$ and assume that the image of $\mathcal{M}$ in $\operatorname{mod}(X)/\mathcal{C}_f$ is simple. Write $e = e(\mathcal{M})$ and $r_{p,n} = r_{p,n}(\mathcal{M})$ for $p \in Y$ and $n \geq 1$. Assume that $a_{p,n}$ are natural numbers such that*

$$\frac{f(f+3)}{2} \geq \sum_{\bar{p},n} \frac{a_{p,n}(a_{p,n}+1)}{2} \tag{10.3}$$

for some integer $f < e$. Then $ef \geq \sum_{\bar{p},n} r_{p,n} a_{p,n}$. Here the notation $\bar{p}$ indicates that we take only one representative from each τ-orbit.

PROOF. The proof consists of several steps.

STEP 1. Assume that $M \in \operatorname{gr}(A)$ contains no finite dimensional submodules and let $\mathcal{M} = \pi M$. Then

$$\dim\{x \in A_n \mid M_0 x = 0\} \geq \dim A_n - \dim \Gamma(X, (\mathcal{O}_X/\operatorname{Ann}\mathcal{M})(n)) \tag{10.4}$$

To prove this we note that the lefthand side of (10.4) is the degree n part of the kernel of the canonical map $A \to M_0^* \otimes_k M$. If we denote this kernel by L then we have to prove

$$\dim_k (A/L)_n \leq \dim \omega(\mathcal{O}_X/\operatorname{Ann}\mathcal{M})_n$$

Now note that there is a canonical map $M_0 \to \Gamma(X, \mathcal{M})$. If we then look at the composition

$$\mathcal{O}_X \to \Gamma(X,\mathcal{M})^* \otimes_k \mathcal{M} \to M_0^* \otimes_k \mathcal{M}$$

we find $\operatorname{Ann}\mathcal{M} \subset \pi L$.

Hence it suffices to prove that

$$\dim_k (A/L)_n \leq \dim_k \omega(\mathcal{O}_X/\pi L)_n = \dim_k (A/L)\tilde{}_n$$

Now A/L is a graded submodule of $M_0^* \otimes_k M$ and hence A/L contains no finite dimensional submodules. Therefore $A/L \hookrightarrow (A/L)\tilde{}$. This proves what we want.

STEP 2. Without loss of generality we may (and we will) assume that $a_{p,n} = 0$ if $r_{p,n} = 0$. Furthermore by definition we have $r_{p,1} \geq r_{p,2} \geq \cdots$ so by permuting the $a_{p,n}$ we may also assume that $a_{p,1} \geq a_{p,2} \geq a_{p,3} \geq$, because in this way (10.3) remains valid and $\sum_{p,n} r_{p,n} a_{p,n}$ does not decrease.

STEP 3. Assume $\underline{a}_p = (a_{p,1}, a_{p,2}, \dots)$ is a non-increasing set of numbers, zero for large n. We associate to such $\underline{a}_p$ an object in $\mathcal{C}_{f,p}$.

Let $(b_{p,n})_{n\geq 1}$ be the partition which is conjugate to $a_{p,n}$. We put $K_p(\underline{a}_p) = \prod_n R/m^{b_{p,-n+1}}$. This definition is to be understood as defining a set of row vectors with the obvious right C_p-action. Then we define $\mathcal{K}_p(\underline{a}_p)$ as the object in $\mathcal{C}_{f,p}$ such that $\hat{\mathcal{K}}_p(\underline{a}_p) = K_p(\underline{a}_p)$.

STEP 4. Below we need the following formula.

$$\dim(C_p/\operatorname{Ann}_{C_p} K_p(\underline{a}_p)) = \sum_n \frac{a_{p,n}(a_{p,n}+1)}{2} \tag{10.5}$$

We leave the obvious verification to the reader.

STEP 5. Now define $\mathcal{K} = \oplus_p \mathcal{K}_p(\underline{a}_p)$ and $K = \omega\mathcal{K}$. For $f < e$ we are going to bound the dimension of

$$\{x \in A_f \mid K_m x = 0\}$$

for large m. This can be done by the method exhibited in Step 1. We find that this dimension is bigger than

$$\dim_k A_f - \dim_k \omega(\mathcal{O}_X/\operatorname{Ann}\mathcal{K})_{m+f}$$

Now

$$\dim_k A_f = \frac{(f+1)(f+2)}{2}$$

and

$$\begin{aligned}
\dim_k \omega(\mathcal{O}_X/\operatorname{Ann}\mathcal{K})_{m+f} &= \operatorname{length}(\mathcal{O}_X/\operatorname{Ann}\mathcal{K}) \\
&= \sum_p \operatorname{length}(\mathcal{O}_X/\operatorname{Ann}\mathcal{K}_p(\underline{a}_p)) \\
&= \sum_p \dim_k(C_p/\operatorname{Ann}_{C_p}(K_p(\underline{a}_p)) \\
&= \sum_{p,n} \frac{a_{p,n}(a_{p,n}+1)}{2}
\end{aligned}$$

where we have used (10.1)(10.5). So ultimately we obtain for large m

$$\dim_k\{x \in A_f \mid K_m x = 0\} \geq \frac{(f+1)(f+2)}{2} - \sum_{p,n} \frac{a_{p,n}(a_{p,n}+1)}{2}$$

STEP 6. Let $\mathcal{M}$ be as in the statement of the proposition. Since $(r_{p,n})_{p,n}$ is unaffected by $\mathcal{C}_f$ we may and we will assume that $\mathcal{M}$ is p-normalized where p runs through a fixed set of representatives for the τ-orbits in Y. Then by lemma 5.7.4 we can recover the structure of $\hat{\mathcal{M}}_p$ from $T_p(\mathcal{M})$. Assume that $T_p(\mathcal{M}) = \oplus_n R/m^{t_{p,n}}$. Then it follows that $\oplus_n K_p(a_{p,1}, \dots, a_{p,t_{p,n}})$ is a quotient of $\hat{\mathcal{M}}_p$.

Hence $\oplus_{p,n}\mathcal{K}_p(a_{p,1}, \dots, a_{p,t_{p,n}})$ is a quotient of $\mathcal{M}$. This yields that $L = \oplus_{p,n}\omega\mathcal{K}_p(a_{p,1}, \dots, a_{p,t_{p,n}})$ is a quotient of $M = \omega\mathcal{M}$ for large m.

STEP 7. Since we have chosen $\mathcal{M}$ to be normalized we have in particular that $\mathcal{M}$ contains no subobject supported on Y. Since the only modules of GK-dimension one of A are coming from objects supported on Y [**9**] it follows that $M = \omega\mathcal{M}$ contains no submodules of GK-dimension one. Since $\mathcal{M}$ was also supposed to be simple modulo $\mathcal{C}_f$ it follows that M is critical. I.e. M contains no non-trivial submodules of multiplicity strictly smaller than e.

If $0 \neq x \in A_f$ then $e(A/xA) = f$. So it follows that $\operatorname{Hom}_A(A/xA, M) = 0$ which is the same as saying that multiplication by x is injective. Since $\operatorname{GKdim} M = 0$ we have for large m: $\dim M_m = em + s$ for some constant s. It follows that for large m one has

$$\dim(M/Mx)_m = \dim M_m - \dim M_{m-f} = (em+s) - (e(m-f)+s) = ef \tag{10.6}$$

Now by hypotheses

$$\frac{f(f+3)}{2} + 1 = \frac{(f+1)(f+2)}{2} > \sum_{p,n} \frac{a_{p,n}(a_{p,n}+1)}{2}$$

Let m be large. By virtue of Step 5 there exists a non-zero $x \in A_f$ such that $K_m x = 0$. Now by construction every indecomposable summand of L is a quotient of K. Thus also $L_m x = 0$.

Now by the Step 6 there is a map $M \to L$, surjective in high degree. Tensoring with A/Ax yields a map $M/Mx \to L/Lx$ with the same property.

We conclude

$$ef = \dim_k (M/Mx)_m \geq \dim_k (L/xL)_m = \dim_k L_m = \sum_{p,n} \dim K_p(a_{p,1}, \dots, a_{p,t_{p,n}})$$

Now an easy verification shows that

$$\sum_n \dim K_p(a_{p,1}, \dots, a_{p,t_{p,n}}) = \sum_n a_{p,n} r_{p,n}$$

which finishes the proof. □

The following proposition would be a trivial consequence of (10.2). With a lot more work we can also deduce it from Proposition 10.1.2.

PROPOSITION 10.1.3. *Assume that $\mathcal{M} \in \operatorname{trans}_Y(X)$ represents a non-zero simple object in $\operatorname{mod}(X)/\mathcal{C}_f$. Let $r = (r_{p,n}(\mathcal{M}))_{p,n}$, $e = e(\mathcal{M})$. If the number of non-zero entries of r is ≤ 6 then there are (up to permutation) only two possibilities for e, r.*

$$e = 1, \qquad r = (1,1,1)$$
$$e = 2, \qquad r = (1,1,1,1,1,1)$$

PROOF. We will apply Proposition 10.1.2. Since this is obviously a numerical criterion we will write $r = (r_i)_{i \in I}$ where i runs through the pairs (p,n) such that $r_{p,n}(\mathcal{M}) \neq 0$. We will consider the entries of r up to permutation. Put $m = |I|$. We will apply Proposition 10.1.2 with suitably chosen $(a_i)_{i \in I}$.

First observe that Proposition 10.1.2 remains true if we allow the $(a_i)_{i \in I}$ to be negative. Indeed if

$$\frac{f(f+3)}{2} \geq \sum_{i \in I} \frac{b_i(b_i+1)}{2}$$

with $b_i \in \mathbb{Z}$ then we define

$$a_i = \begin{cases} b_i & \text{if } b_i \geq 0 \\ -b_i - 1 & \text{if } b_i < 0 \end{cases}$$

Then $\sum_i b_i(b_i+1)/2 = \sum_i a_i(a_i+1)/2$ and $a_i \geq b_i$. By Proposition 10.1.2 we have $ef \geq \sum_i r_i a_i$ and thus $ef \geq \sum_i r_i b_i$.

Now we can reformulate Proposition 10.1.2 as follows: the intersection of the closed ball

$$B = \{(a_i)_{i\in I} \mid \sum \frac{a_i(a_i+1)}{2} \leq \frac{f(f+3)}{2}\}$$

and the open half space

$$H = \{(a_i)_{i\in I} \mid \sum_i r_i a_i > ef\}$$

contains no integral point.

Now $B \cap H$ certainly contains an open ball with diameter equal to the radius of B minus the distance of H to the center of B.

The center of B is $(-\frac{1}{2}, \dots, -\frac{1}{2})$ and the radius of B is given by

$$\sqrt{f(f+3) + \frac{m}{4}}$$

The distance of H to the center of B is given by

$$\frac{ef + \frac{\sum_i r_i}{2}}{\|r\|} \tag{10.7}$$

where $\|r\| = \sqrt{\sum_i r_i^2}$.

If we use the fact that $3e = \sum_i r_i$ then (10.7) becomes equal to

$$\frac{e(f+\frac{3}{2})}{2}$$

Thus $B \cap H$ contains an open ball of diameter

$$\sqrt{f(f+3) + \frac{m}{4}} - \frac{e(f+\frac{3}{2})}{\|r\|}$$

Now it is easy to see that an open ball of diameter $> \sqrt{m}$ must contain a point with integral coordinates. Thus Proposition 10.1.2 yields

$$\sqrt{f(f+3) + \frac{m}{4}} - \frac{e(f+\frac{3}{2})}{\|r\|} \leq \sqrt{m} \tag{10.8}$$

From $\sum_i r_i = 3e$ we obtain the estimate $\|r\| \geq 3e/\sqrt{m}$. Combining this with (10.8) yields

$$\sqrt{f(f+3) + \frac{m}{4}} \leq \left(\frac{f}{3} + \frac{3}{2}\right)\sqrt{m}$$

We now take $f = e - 1$. Then

$$(e-1)(e+2) + \frac{m}{4} \leq \left(\frac{e}{3} + \frac{7}{6}\right)^2 m$$

which yields

$$\frac{(e-1)(e+2)}{\left(\frac{e}{3}+\frac{7}{6}\right)^2 - \frac{1}{4}} \leq m$$

If we combine this with the hypotheses $m \leq 6$ then we obtain $e \leq 13$.

Explicit enumeration of all possibilities for e, r (using a computer) yields that $B \cap H$ always contains an integral point for $e \geq 3$. So the remaining possibilities are $e = 1, 2$.

If $e = 2$ then if follows easily from Proposition 10.1.2 that $r = (1, 1, 1, 1, 1, 1)$. However if $e = 1$ then Proposition 10.1.2 gives no information whatsoever. But is $e(\mathcal{M}) = 1$ then $\mathcal{M} = \pi M$ where M is a so-called "line-module" [**9**]. We will analyze this case directly.

According to Proposition 5.7.2 we may change $\mathcal{M}$ in such a way that $\mathcal{M}$ is p normal, where p runs through a set of representatives of the τ-orbits on Y. Then we can read of r from the decomposition of $\mathcal{M}/\mathcal{M}(-Y)$ into uniserial $\mathcal{O}_Y$-modules.

As indicated above $\mathcal{M} = A/xA$ where $x \in A_1$. But then $\mathcal{M}/\mathcal{M}(-Y)$ is given by the zeroes of $x \in \Gamma(Y, \mathcal{L})$. Since $\mathcal{L}$ defines an embedding of Y in $\mathbb{P}^2$ this can be interpreted as the scheme theoretic intersection of Y and $V(x)$. This is a union of uniserial schemes, such that every point in Y occurs at most once in the reduced locus. Hence we obtain that $r = (1, 1, 1)$. □

10.2. Classification of lines and conics

The cases $e(\mathcal{M}) = 1$ and $e(\mathcal{M}) = 2$ represent lines and conics. The first type of object has been studied in [**9**] and the second type of object has been studied in [**3**].

If $\mathcal{M} \in \operatorname{trans}_Y(X)$ then let us denote by $\overline{\operatorname{Div}}(\mathcal{M})$ the image of $\operatorname{Div}(\mathcal{M})$ in $\mathbb{N}(Y/\langle\tau\rangle)$ under the map which sends $p \in Y$ to its τ-orbit.

Let us call $M \in \operatorname{Gr}(A)$ *standard* if $M_0 \neq 0$ and $M_i = 0$ if $i < 0$. Let us call $M \in \operatorname{gr}(A)$ Cohen-Macaulay of dimension d if M has GK-dimension d and if $\operatorname{Ext}^s_A(M, A) = 0$ unless $s = 3 - d$. The following is easy to see and will be used below (cfr [**2**]).

LEMMA 10.2.1. *Assume that $M \in \operatorname{gr}(A)$ has GK-dimension two and that M contains no submodules of GK-dimension one. Then $\tilde{M}$ is Cohen-Macaulay.*

According to [**3**] and [**9**] there are the following three classes of critical Cohen-Macaulay modules of GK-dimension two and multiplicity 1 or 2.

1. Lines: modules of the form A/xA, $x \in A_1 - \{0\}$.
2. Conics of the first kind: critical graded A-modules of the form A/xA, $x \in A_2 - \{0\}$.
3. Conics of the second kind: critical graded A-modules with minimal resolution

$$0 \to A(-1)^2 \to A^2 \to M \to 0$$

Shifts of such modules will be called shifted lines and conics (of the first and the second kind). They exhaust all critical Cohen-Macaulay modules of GK-dimension two and multiplicity ≤ 2 [**3**].

Below L will be an effective divisor on Y such that $\mathcal{L} = \mathcal{O}_Y(L)$. The following results from [**3**] describe the divisor classes of modules of multiplicity ≤ 2.

PROPOSITION 10.2.2. 1. *If $M = N(s)$ where N is a line module then $\operatorname{Div} \pi M$ is of the form $\sigma^s L$ for some s.*
2. *If $M = Q(s)$ where Q is a conic of the first kind then $\operatorname{Div} \pi M \sim \sigma^s L + \sigma^{s-1} L$.*
3. *If $M = Q(s)$ where Q is a conic of the second kind then $\operatorname{Div} \pi M \sim \sigma^s L + \sigma^s L$.*

If $D \in \mathbb{N}(Y/\langle\tau\rangle)$ and H is a divisor on E then we say that D is compatible with H if there are $(p_i)_{i=1,\dots,l}$ with $l = \deg H$ such that $D = \sum_{i=1}^{l} \overline{p}_i$ and $H \sim \sum_{i=1}^{l} p_i$.

From the previous proposition we deduce the following:

LEMMA 10.2.3. *Assume that M is a Cohen-Macaulay module of GK-dimension 2. If $e(M) = 1$ then $\overline{\mathrm{Div}}(\pi M)$ is compatible with L. If $e(M) = 2$ then $\overline{\mathrm{Div}}(\pi M)$ is compatible with $2L$.*

PROOF. This follows from the fact that the image of $\sigma^s L$ in $\mathbb{N}(Y/\langle\tau\rangle)$ is compatible with L. This follows in turn from the fact that $\tau = 3\sigma$ and $\deg L = 3$. □

We also deduce the following lemma.

LEMMA 10.2.4. *Assume that $M \in \mathrm{gr}(A)$ is a Cohen-Macaulay module of GK-dimension 2. Then the fact whether it is a shifted conic of the first or second kind can be recognized from $\mathrm{Div}\,\pi M$.*

On the other hand we can't recognize the kind of a conic from its image in $\mathrm{trans}_Y(X)$. In fact we have the following easy lemma.

LEMMA 10.2.5. *Assume that $\mathcal{N} \in \mathrm{trans}_Y(X)$ is such that $e(\mathcal{N}) = 2$. Then $\mathcal{N}$ is equivalent modulo $\mathcal{C}_f$ to an object of the form $\pi(A/xA)(m)$.*

PROOF. We may assume that $\mathcal{N}$ is Y-torsion free. Let $N = \omega(\mathcal{N})$. If N is a shifted conic of the first kind then we are done. Assume that this is not the case. Let p be a point in the support of $\mathrm{Div}(\mathcal{N})$ and let $\mathcal{N}'$ be the kernel of the associated map $\mathcal{N} \to \mathcal{O}_p$. Put $N' = \omega\mathcal{N}'$. Then by the formula (5.37) together with the discussion preceding lemma 10.2.4 we deduce that N' is of the first kind. □

THEOREM 10.2.6. *The map $\overline{\mathrm{Div}}$ defines a bijection between*

1. *simple objects in $\mathrm{trans}_Y(X)/\mathcal{C}_f$ of multiplicity 1 and elements of $\mathbb{N}(Y/\langle\tau\rangle)$ compatible with L;*
2. *simple objects in $\mathrm{trans}_Y(X)/\mathcal{C}_f$ of multiplicity 2 and elements of $\mathbb{N}(Y/\langle\tau\rangle)$ compatible with $2L$ that are not the sum of two elements of $\mathbb{N}(Y/\langle\tau\rangle)$ compatible with L.*

For the proof we need the following easy lemma.

LEMMA 10.2.7. *Assume that $\mathcal{M} \in \mathrm{trans}_Y(X)$ is Y-torsion free and assume that $\mathrm{Div}(\mathcal{M}) = \sum_{i=1}^{l} p_i$ where the $(q_i)_i$ are in the τ-orbit of the $(p_i)_i$. Then there exists $\mathcal{N} \in \mathrm{trans}_Y(X)$ which is Y-torsion free and which is equivalent to $\mathcal{M}$ modulo $\mathcal{C}_f$ such that $\mathrm{Div}(N) = \sum_i q_i$.*

PROOF. To prove this we first replace $\mathcal{M}$ by $\mathcal{M}(mY)$, m large and then apply the formula (5.37). □

PROOF OF THEOREM 10.2.6. Note that by lemma 10.2.3 if $\mathcal{M} \in \mathrm{trans}_Y(X)$ then $\overline{\mathrm{Div}}(\mathcal{M})$ is compatible with L or $2L$ if $e(\mathcal{M})$ is 1 or 2.

We first prove that $\overline{\mathrm{Div}}$ is a bijection between simple objects in $\mathrm{trans}_Y(X)/\mathcal{C}_f$ of multiplicity 1 and elements of $\mathbb{N}(Y/\langle\tau\rangle)$ compatible with L. Surjectivity is obvious so we consider injectivity.

Assume that $\mathcal{N}_{1,2} \in \mathrm{trans}_Y(X)$ have multiplicity 2 and $\overline{\mathrm{Div}}(\mathcal{N}_1) = \overline{\mathrm{Div}}(\mathcal{N}_2)$. According to lemma 10.2.7 and the previous discussion we can assume that $\mathcal{N}_1 = \pi(A/xA)(m)$, $\mathcal{N}_2 = (A/yA)(n)$ in such a way that $\mathrm{Div}(\mathcal{N}_1) = \mathrm{Div}(\mathcal{N}_2)$. Now consider x, y as global sections of $\mathcal{L}$ on Y. Then $\operatorname{Div} \mathcal{N}_1 = \sigma^m \operatorname{div}(x)$, $\operatorname{Div} \mathcal{N}_2 = \sigma^n \operatorname{div}(y)$. Since σ has infinite order this implies $m = n$ and $x = y$ (up to a scalar).

Now we prove the second part of the theorem. We consider injectivity first. Assume that $\mathcal{N}_{1,2} \in \operatorname{trans}_Y(X)$ have multiplicity $= 2$ and are such that $\overline{\operatorname{Div}}(\mathcal{N}_1) = \overline{\operatorname{Div}}(\mathcal{N}_2)$. Using lemmas 10.2.5,10.2.7 we may assume that $\mathcal{N}_1 = \pi(A/xA)(m)$ and $\operatorname{Div}(\mathcal{N}_1) = \operatorname{Div}(\mathcal{N}_2)$, where $\mathcal{N}_2$ is in addition Y-torsion free. But then by lemma 10.2.4 we find that $\mathcal{N}_2$ is obtained from a shifted conic of the first kind, whence $\mathcal{N}_2 = \pi((A/yA)(n))$. Now exactly as in the case $e = 1$ this implies $x = y$, $m = n$.

Now we have to describe the image of $\overline{\operatorname{Div}}$. First let $D \in \mathbb{N}(Y/\langle\tau\rangle)$ be compatible with $2L$ and not be the sum of two elements compatible with L. Then according to lemma 10.2.3 if $\overline{\operatorname{Div}}(A/xA) = D$ then A/xA is critical.

Assume on the other hand that $\mathcal{M} \in \operatorname{trans}_Y(X)$ is such that $e(\mathcal{M}) = 2$ and $\overline{\operatorname{Div}}(\mathcal{M}) = D_1 + D_2$, $D_{1,2}$ compatible with L. We claim that $\mathcal{M}$ is not simple modulo $\mathcal{C}_f$. Without loss of generality we may assume that $\mathcal{M}$ is Y-torsion free.

Using lemma 10.2.7 we may assume that $\operatorname{Div} \mathcal{M} = E_1 + E_2$ with $E_1 \sim L$, $E_2 \sim \sigma^{-1} L$. Put $M = \omega\mathcal{M}$.

It follows from Proposition 10.2.2 that $M = A/xA$ with $x \in A_2 - \{0\}$. Let N be the line module corresponding to E_1. Since the divisor of x contains E_1 it follows that A/xA maps surjectively to N/Ng. Since $x \in A_2$ this implies that M maps surjectively to N. Hence A/xA is not critical. $\square$

CHAPTER 11

Blowing up n points in an elliptic quantum plane

11.1. Derived categories

In this section our notations and conventions will be as in Section §10 except that we do *not* assume that σ has infinite order.

We choose n points $p_1, \ldots, p_n \in Y$ where $n \leq 8$ (at this point not necessarily distinct). We define quasi-schemes X_j, $j = 1, \ldots, n+1$, $\tilde{X}_j$, $j = 1, \ldots, n$ containing Y as a divisor. We do this as follows :

1. $X_1 = X$.
2. $\tilde{X}_j$ is the blowup of X_j at the point p_j.
3. X_{j+1} is derived from the pair (X_j, p_j) in the same way as V was derived from (X, p) in §9.2.

Of course we are only allowed to make this construction if Hypothesis (****) of §9.2 holds for the triples (Y, X_j, p_j). Let us check this now by induction.

Smoothness of the p_j is by hypotheses.

Ampleness of $o_{X_j}(Y)$ for $j \geq 2$ follows from the fact that X_j is the Proj of a graded algebra which satisfies χ (Proposition 9.2.1) and Y is defined by a central element of degree one. This argument also works for X_1 if we note that $X_1 = \operatorname{Proj} A^{(3)}$ where $A^{(3)}$ denotes the 3-Veronese of A.

Hence the only thing that remains to be checked is that I_{Y,p_j} is ample (we have now included the point p_j in the prior notation I_Y). We check ampleness by induction. Assume that I_{Y,p_t} is ample for $t < i$. By definition we have $I_{Y,p_j} = m_{Y,p_j} \mathcal{N}_{Y/X_j}$. By (9.5) we have that $\mathcal{N}_{Y/X_j} = \mathcal{N}_{Y/\tilde{X}_{j-1}}$ and by (6.26) we have $\mathcal{N}_{Y/\tilde{X}_{j-1}} = m_{p_{j-1}} \mathcal{N}_{Y/X_{j-1}}$.

Since $\mathcal{N}_{Y/X}$ is obtained by the functor $- \otimes Ag^{-1}$ and $Ag^{-1} = A(3)$ one deduces from **[10]** that

$$\mathcal{N}_{Y/X} = (\mathcal{L}_\sigma)^{\otimes 3} = (\mathcal{L} \otimes_{\mathcal{O}_Y} \sigma^*(\mathcal{L}) \otimes_{\mathcal{O}_Y} \sigma^{2*}(\mathcal{L}))_{\sigma^3}$$

Thus $\mathcal{N}_{Y/X} = \mathcal{N}_\tau$ with $\tau = \sigma^3$ and $\mathcal{N} = (\mathcal{L} \otimes_{\mathcal{O}_Y} \sigma^*(\mathcal{L}) \otimes_{\mathcal{O}_Y} \sigma^{2*}(\mathcal{L}))_{\sigma^3}$. In particular $\deg \mathcal{N} = 9$.

Define

$$\mathcal{N}_j = (m_{Y,p_1} \cdots m_{Y,p_{j-1}} \mathcal{N})_\tau \tag{11.1}$$

By the above discussion we obtain $\mathcal{N}_{Y/X_j} = \mathcal{N}_{Y/\tilde{X}_{j-1}} = (\mathcal{N}_j)_\tau$ and $I_{Y,p_j} = (\mathcal{N}_{j+1})_\tau$.

Since $\deg \mathcal{N}_j = 9 - j + 1$ we obtain that $\mathcal{N}_{j+1}$ has positive degree if $j \leq 9$. This is true by the restriction $n \geq 8$. Hence I_{Y,p_j} is ample.

With our current notations the following diagram replaces (9.17)

$$\begin{array}{ccccc} & & Y & & \\ & \swarrow{\scriptstyle i} & \downarrow{\scriptstyle i} & \searrow{\scriptstyle i} & \\ X_j & \xleftarrow[\alpha_j]{} & \tilde{X}_j & \xrightarrow[\delta_j]{} & X_{j+1}\end{array} \tag{11.2}$$

We will denote the exceptional curves in $\tilde{X}_j$ by $\mathcal{O}_{L_j}$ and their direct images in X_{j+1} by $\mathcal{O}_{M_{j+1}}$.

LEMMA 11.1.1. *The hypotheses for Proposition 9.2.3 hold for X_j, Y, p_j. In particular we have :*

1. $R\Gamma(X_j, \mathcal{O}_{X_j}) = R\Gamma(\tilde{X}_j, \mathcal{O}_{\tilde{X}_j}) = k$.
2. $R\delta_{j,*}\, L\delta_j^*$ *is the identity on* $D^-(X_{j+1})$.

PROOF. The fact that the hypotheses for Proposition 9.2.3 hold is verified by induction, starting from the easy fact that $R\Gamma(X, \mathcal{O}_X) = k$. □

We borrow the following definition from the commutative case.

DEFINITION 11.1.2. Assume that $(q_j)_{j=1,\dots,n}$, $n \leq 8$ are points in $\mathbb{P}^2$. The points $(q_j)_j$ are in general position if and only if

- All points are different.
- No three points lie on a line.
- No six points lie on a conic.
- If $n = 8$ then not all points lie on a singular cubic divisor in such a way that one of the points lies on the singularity.

The following will be our main theorem.

THEOREM 11.1.3. *Assume that the points $(\sigma p_j)_j$ are in general position (with respect to the embedding of $Y \subset \mathbb{P}^2$, fixed in the beginning of this section). Then the following holds for all j :*

1. δ_{j*}, δ_j^* *have cohomological dimension* ≤ 1.
2. $L\delta_j^*$, $R\delta_{j*}$ *are inverse equivalences between* $D(\tilde{X}_j)$ *and* $D(X_{j+1})$.
3. $\mathrm{Mod}(X_j)$, $\mathrm{Mod}(\tilde{X}_j)$ *have finite injective dimension.*

PROOF. Without loss of generality we may assume that $j = n$. 1.,2. of the theorem will be proved directly but for 3. we will need induction. That is, we will assume that $\mathrm{Mod}(X_j)$ has finite injective dimension for $j < n+1$. Note that this is clearly true if $j = 1$ since X_1 is the Proj of a graded algebra of finite global dimenion. By Corollary (8.4.3) we find that $\mathrm{Mod}(\tilde{X}_j)$ also has finite injective dimension for $j < n+1$.

We already know by lemma 11.1.1 that $R\delta_{n,*}\, L\delta_n^*$ is the identity on $D^-(X_{n+1})$. We will start by showing that $L\delta_n^*\, R\delta_{n,*}$ is the identity on $D_f^-(\tilde{X}_n)$. By lemma 8.1.4 this means that we have to show that $\ker R\delta_{n,*} = 0$. Hence assume that $\mathcal{M} \in D_f^-(\tilde{X}_n)$ is such that $R\delta_{n,*}\mathcal{M} = 0$. We will construct $\mathcal{Q}_j \in D_f^-(X_j)$ for $j = 1, \dots, n$ in such a way that the following holds :

(a) $R\Gamma(X_j, \mathcal{Q}_j) = 0$ (note that $R\Gamma(X_j, -)$ is well defined on $D^-(X_j)$ because $\mathrm{Mod}(X_j)$ has finite injective dimension by the induction hypotheses).

(b) There are triangles

$$\begin{array}{ccc} & \mathcal{U}_n & \\ \swarrow & & \nwarrow \\ L\alpha_n^* \mathcal{Q}_n & \longrightarrow & \mathcal{M} \end{array} \tag{11.3}$$

$$\begin{array}{ccc} & \mathcal{U}_j & \\ \swarrow & & \nwarrow \\ L\alpha_j^* \mathcal{Q}_j & \longrightarrow & L\delta_j^* \mathcal{Q}_{j+1} \end{array} \quad \text{for } j = 1, \ldots n-1 \tag{11.4}$$

where the $\mathcal{U}_j$ are direct sums of shifts in the derived category of copies of $\mathcal{O}_{L_j}(-Y)$.

To do this we define

$$\mathcal{Q}_n = R\alpha_{n,*}\mathcal{M}$$
$$\mathcal{Q}_j = R\alpha_{j,*}L\delta_j^*\mathcal{Q}_{j+1}$$

By adjointness we have maps

$$L\alpha_n^*\mathcal{Q}_n \to \mathcal{M}$$
$$L\alpha_j^*\mathcal{Q}_j \to L\delta_j^*\mathcal{Q}_{j+1}$$

which become the identity after applying $R\alpha_{j,*}$, $j = 1, \ldots, n$. The existence of the triangles (11.3)(11.4) now follows from Theorem 8.4.1. By admissibility we also have

$$R\Gamma(X_n, \mathcal{Q}_n) = R\Gamma(\tilde{X}_n, \mathcal{M}) = R\Gamma(X_{n+1}, R\delta_{*,n}\mathcal{M}) = 0$$

and

$$R\Gamma(X_j, \mathcal{Q}_j) = R\Gamma(\tilde{X}_j, L\delta_j^*\mathcal{Q}_{j+1}) = R\Gamma(X_{j+1}, \mathcal{Q}_{j+1})$$

(for the last equality we have used lemma 11.1.1.2). Hence by induction $R\Gamma(X_j, \mathcal{Q}_j)$ is equal to zero. This finishes the proof of (a) and (b) above.

To continue we use restriction to Y. Applying Li^* to (11.3) and (11.4) yields triangles

$$\begin{array}{ccc} & Li^*\mathcal{U}_n & \\ \swarrow & & \nwarrow \\ Li^* \mathcal{Q}_n & \longrightarrow & Li^*\mathcal{M} \end{array} \tag{11.5}$$

$$\begin{array}{ccc} & Li^*\mathcal{U}_j & \\ \swarrow & & \nwarrow \\ Li^* \mathcal{Q}_j & \longrightarrow & Li^* \mathcal{Q}_{j+1} \end{array} \quad \text{for } j = 1, \ldots n-1 \tag{11.6}$$

where the $Li^*\mathcal{U}_j$ are direct sums of shifts in the derived category of $\mathcal{O}_{\tau p_j}(-Y) = \mathcal{O}_{p_j}$.

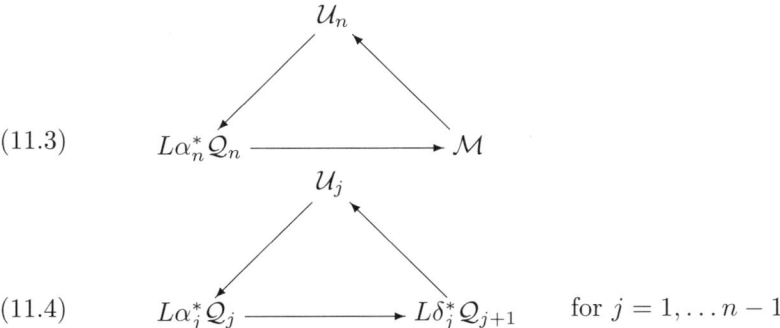

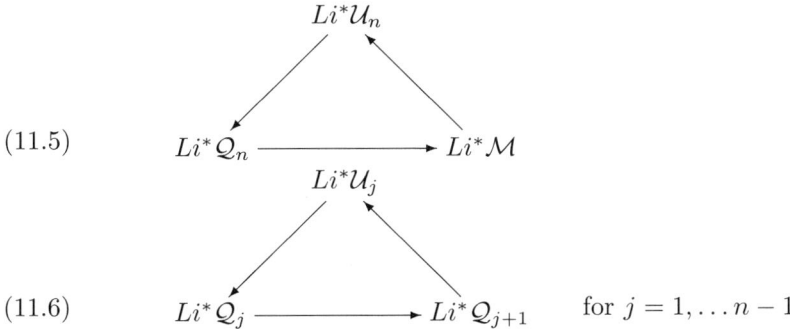

Now by hypotheses $R\delta_{*,n}\mathcal{M} = 0$. Thus by (9.12) we obtain $0 = Li^*R\delta_{*,n}\mathcal{M} = Li^*\mathcal{M}$. Hence $Li^*\mathcal{Q}_n = Li^*\mathcal{U}_n$.

Taking into account that $\mathrm{RHom}_{D(Y)}(\mathcal{O}_{p_j}, \mathcal{O}_{p_t}) = 0$ if $j \neq t$ we deduce that the map $Li^*\mathcal{Q}_n \to Li^*\mathcal{U}_{n-1}$, must be the zero map in (11.6). Hence $Li^*\mathcal{Q}_{n-1} = Li^*\mathcal{U}_{n-1}[-1] \oplus Li^*\mathcal{Q}_n = Li^*\mathcal{U}_{n-1}[-1] \oplus Li^*\mathcal{U}_n$. Continuing yields

$$Li^*\mathcal{Q}_1 = Li^*\mathcal{U}_1[-n+1] \oplus \cdots \oplus Li^*\mathcal{U}_n \tag{11.7}$$

In particular $Li^*\mathcal{Q}_1$ is a direct sum of shifts in the derived category of copies of $\mathcal{O}_{p_j}$.

Now $\mathcal{Q}_1$ lives on $X_1 = X$ and this is a very well understood quasi-scheme. In fact it follows from lemma 11.1.4 below that necessarily $\mathcal{Q}_1 = 0$. But then from (11.7) we immediately deduce that $\mathcal{U}_j = 0$.

From the triangle (11.4) we then find that $L\delta_1^*\mathcal{Q}_2 = 0$. Applying $R\delta_{*,1}$, using lemma 11.1.1 yields that $\mathcal{Q}_2 = 0$. Continuing by induction we eventually find that $\mathcal{Q}_n = 0$, and hence by (11.3) we finally obtain $\mathcal{M} = 0$.

At this point we have partially proved 2. Let us now prove that δ_n^* has cohomological dimension ≤ 1. Let $\mathcal{M} \in \mathrm{Mod}(X_n)$ and put $\mathcal{N} = L\delta_n^*\mathcal{M}$. From the spectral sequence

$$R^p\delta_{n,*}H^q(\mathcal{N}) \to R^{p+q}\delta_{n,*}\mathcal{N}$$

we obtain exact sequences

$$0 \to R^1\delta_{n,*}H^{i-1}(\mathcal{N}) \to R^i\delta_{n,*}\mathcal{N} \to R^0\delta_{n,*}H^i(\mathcal{N}) \to 0 \tag{11.8}$$

Since $R^i\delta_{n,*}\mathcal{N} = R^i\delta_{n,*}L\delta_n^*\mathcal{M} = H^i(\mathcal{M})$ we deduce that $R^i\delta_{n,*}\mathcal{N} = 0$ for $i < 0$. By (11.8) this yields that $R^1\delta_{n,*}H^i(\mathcal{N}) = 0$ for $i < -1$ and $R^0\delta_{n,*}H^i(\mathcal{N})$ for $i < 0$. In particular $R\delta_{n,*}H^i(\mathcal{N}) = 0$ for $i < -1$. By the part of 2. already proved, this implies $H^i(\mathcal{N}) =$ for $i < -1$. Hence $\mathrm{cd}\,\delta_n^* \leq 1$. This finishes the proof of 1.

Since it now follows that $L\delta_n^*$ is defined on the unbounded derived category, we can prove 2. completely. We have to show that the adjunction mappings are isomorphisms on the unbounded derived category. Take for example $\mathcal{M} \in D(\tilde{X}_n)$. We have to show that $\mathrm{cone}(L\delta_n^*R\delta_{n,*}\mathcal{M} \to \mathcal{M})$ is zero. In the same way as in Step 6 of the proof of Theorem 8.4.1 we reduce to the case $\mathcal{M} \in \mathrm{mod}(\tilde{X}_n)$. But then $\mathcal{M} \in D_f^-(\tilde{X}_n)$ and this case was already handled.

The proof that the other adjunction morphism is an isomorphism is entirely similar.

By the induction hypotheses (and the discussion in the first paragraph of this proof) we know that $\mathrm{Mod}(\tilde{X}_n)$ has finite injective dimension. By an argument similar to the proof of Corollary 8.4.3 we then find that $\mathrm{Mod}(X_{n+1})$ has finite injective dimension (using 1. and 2.). This finishes the proof. $\square$

LEMMA 11.1.4. *Let Y, X and (p_j) be as above. Assume that the σp_j are in general position. Let $\mathcal{T} \in D_f^-(X)$ be an object such that*

1. $R\Gamma(X, \mathcal{T}) = 0$
2. *The homology of $Li^*\mathcal{T}$ is a direct sum of copies of $\mathcal{O}_{p_j}$.*

Then $\mathcal{T} = 0$.

PROOF. Define $\mathcal{E} = \mathcal{O}_X(-2) \oplus \mathcal{O}_X(-1) \oplus \mathcal{O}_X$ and $H = \operatorname{End}_{o_X}(\mathcal{E})$. Thus

$$H = \begin{pmatrix} k & 0 & 0 \\ A_1 & k & 0 \\ A_2 & A_1 & k \end{pmatrix}$$

By a standard generalization of [**11**] it follows that the functor

$$F : D_f^-(X) \to D_f^-(H) : \mathcal{T} \mapsto \operatorname{RHom}(\mathcal{E}, \mathcal{T}) \tag{11.9}$$

is an equivalence.

Let e_i, $i = 1, 2, 3$ be the diagonal idempotents in H and let $P_i = e_i H$ be the corresponding projectives. Under F we have the following correspondences

$$\mathcal{O}_X(-2) \leftrightarrow P_1 \qquad \mathcal{O}_X(-1) \leftrightarrow P_2 \qquad \mathcal{O}_X \leftrightarrow P_3$$

From (11.9) we deduce that

$$R\Gamma(X, \mathcal{T}) = 0 \iff (F\mathcal{T})e_3 = 0$$

Put $H' = (1 - e_3)H(1 - e_3) = \left(\begin{smallmatrix} k & 0 \\ A_1 & k \end{smallmatrix}\right)$.

Right modules over H can be written in block form as row vectors (M_1, M_2, M_3). Similarly H'-modules can be written as (M_1, M_2). Sending $(M_1, M_2) \mapsto (M_1, M_2, 0)$ defines an exact functor $I : \operatorname{Mod}(H') \to \operatorname{Mod}(H)$ which extends to an exact functor $I : D_f^-(H') \to D_f^-(H)$. It is easy to see that this functor defines an equivalence of $D_f^-(H')$ with the full subcategory of $D_f^-(H)$ consisting of objects T such that $Te_3 = 0$. Hence we find in particular that $\ker R\Gamma(X, -)$ is equivalent to $D_f^-(H')$.

Now H' is hereditary, and hence by lemma 8.3.4 an element of $D_f^-(H')$ is the sum of its homology. Assume that $T \in \operatorname{mod}(H')$. Then T has a minimal resolution of length 2

$$0 \to W \otimes_k P_1 \to U \otimes_k P_2 \oplus V \otimes_k P_1 \to T \to 0$$

The complex $W \otimes_k P_1 \to U \otimes_k P_2 \oplus V \otimes_k P_1$ splits as a direct sum of $W \otimes_k P_1 \to U \otimes_k P_2$ and $V \otimes_k P_1$.

Now we view T as a H-module via the functor I defined above. We find that $F^{-1}T$ is a sum of

$$W \otimes_k \mathcal{O}_X(-2) \to U \otimes_k \mathcal{O}_X(-1) \tag{11.10}$$

and $V \otimes_k \mathcal{O}_X(-2)$. We conclude that $\mathcal{T}$ is a direct sum of shifts in the derived category of complexes of the form (11.10) and of complexes of the form $\mathcal{O}_X(-2)$.

By hypotheses we know in addition that the homology of $Li^*\mathcal{T}$ is given by direct sums of copies of $\mathcal{O}_{p_j}$. This yields that $Li^*\mathcal{T}$ must be a direct sum of complexes

$$W \otimes_k \mathcal{O}_Y(-2) \xrightarrow{\phi} U \otimes_k \mathcal{O}_Y(-1)$$

where ϕ is monic and $\operatorname{coker} \phi$ is a direct sum of copies of $\mathcal{O}_{p_j}$. In particular $\dim W = \dim U$.

For convenience we tensor (11.10) with $o_Y(2)$. Taking into account that $\mathcal{O}_Y(1)$ is equal to $\sigma_*(\mathcal{L})$ we obtain a complex

$$W \otimes_k \mathcal{O}_Y \to U \otimes_k \sigma_*(\mathcal{L})$$

whose cokernel which is a direct sum of copies of $\mathcal{O}_{p_j}(2) = \mathcal{O}_{\sigma^2 p_j}$.

Now we invoke Lemma 11.1.6 below with $\mathcal{M} = \sigma_*(\mathcal{L})$ and $q_j = \sigma^2 p_j$. By hypotheses the q_j are in general position with respect to the embedding defined by $\mathcal{M}$. We obtain $W = U = 0$ and the proof is done. $\square$

The above proof was based on lemma 11.1.6 below. To prove this lemma we need some notions of the theory of commutative blowing up.

Assume that Z is a smooth (commutative!) surface let $Y \subset Z$ be a divisor. Let $p \in Y$ be a smooth point and let $\psi : \tilde{Z} \to Z$ and $\tilde{Y}$ be respectively the blowup of Z at p and the strict transform of Y. As usual $\tilde{Z} = \mathrm{Proj}(\oplus_n m_p^n)$, where m_p is the maximal ideal of $\mathcal{O}_Z$ defined by p. Let $L \subset \tilde{Z}$ be the exceptional curve.

Let $\mathcal{P}$ be a one-dimensonal coherent Cohen-Macaulay module on Z. We define the *strict transform* $\psi^{-1}(\mathcal{P})$ of $\mathcal{P}$ as $\pi(\oplus_n m_p^n \mathcal{P})$. The following lemma is easily proved.

LEMMA 11.1.5. *Let the notation be as above. Assume that Y is not contained in the support of $\mathcal{P}$ and furthermore that the p-primary component of $i^*\mathcal{P}$ is of the form $\mathcal{O}_p^{\oplus u}$. Then* $\mathrm{Supp}(\psi^{-1}\mathcal{P}) \cap \tilde{Y} \cap L = \emptyset$.

We use this lemma to prove the following result.

LEMMA 11.1.6. *Let Y be embedded as a cubic divisor in $\mathbb{P}^2$ through a very ample line bundle $\mathcal{M}$ of degree three. Assume that $(q_j)_{j=1,\dots,n} \in Y$, $n \leq 8$, are smooth points in general position. Then it is impossible to have a map $\mathcal{O}_Y^u \xrightarrow{\phi} \mathcal{M}^u$ whose cokernel is a direct sum of copies of $\mathcal{O}_{q_j}$, unless $u = 0$.*

PROOF. Let $t : Y \to \mathbb{P}^2$ be the embedding. The map ϕ can be uniquely lifted to a map $\mathcal{O}_{\mathbb{P}^2}^u \xrightarrow{\mu} \mathcal{O}_{\mathbb{P}^2}^u$ such that $\phi = t^*\mu$. Let $\mathcal{P} = \operatorname{coker} \mu$. Then $\mathcal{P}$ is a one dimensional Cohen-Macaulay module on $\mathbb{P}^2$ such that $t^*\mathcal{P}$ is a direct sum of copies of $\mathcal{O}_{q_j}$. We claim that this is impossible unless $\mathcal{P} = 0$.

To do this we perform the blowup of $\mathbb{P}^2$ at $q_1, \dots, q_n$. Let $\tilde{\mathcal{P}}$ and $\tilde{Y}$ be the (iterated) strict transforms of $\mathcal{P}$ and Y. Then according to lemma 11.1.5 we have $\operatorname{Supp} \tilde{\mathcal{P}} \cap \tilde{Y} = \emptyset$. On the other hand it follows from the Nakai criterion that if the q_j are in general position then $\tilde{Y}$ is ample. This is clearly a contradiction. □

11.2. Exceptional simple objects

If $Y \subset X$ is a commutative curve contained as a divisor in a quasi-scheme X then we will call a simple object in $\mathrm{mod}(X)$ *exceptional* if it is not of the form $\mathcal{O}_p$ for $p \in Y$. One of the aims of these notes is to count the exceptional simple objects in the quasi-schemes X_n which were introduced in the previous section. So far we have only been able to do this under some additional hypotheses as can be seen from our main result below (Theorem 11.2.1). In this section we use the same notations and hypothese as in sections §10 and §11.1. We assume in addition that τ has infinite order.

Throughout we will choose a group law on Y in such a way that if $p, q, r \in Y$ lie on a line then $p + q + r = 0$. Furthermore we choose a fixed set $F \subset E$ of representatives for the τ-orbits on Y. We partially order $\mathbb{N}F$ by putting $y \leq z$ if $y_p \leq z_p$ for all $p \in F$. If $z \in \mathbb{N}F$ then we write $|z| = \sum_{p\in F} z_p$, $N(z) = \sum_{p \in F} z_p p$.

For $n \in \mathbb{N}$ we define

$$H_n = \{y \in \mathbb{N}F \mid |y| = n, N(y) \in \mathbb{Z}\tau\}$$

and for $z \in \mathbb{N}F$ we also define

$$A_z = \{y \in H_3 \mid y \leq z\}$$

$$B_z = \{y \in H_6 \mid y \leq z, y \text{ is not the sum of two elements of } H_3\}$$

We now have the following theorem.

THEOREM 11.2.1. *Let* $p_1, \dots, p_n \in Y$, $n \leq 6$ *be such that* $(\sigma p_i)_i$ *are in general position (Def. 11.1.2). For* $p \in F$ *let* z_p *be the cardinality of the intersection of* $\{p_1, \dots, p_n\}$ *with the* τ*-orbit of* p. *Let* O *be the number of non-zero* z_p. *Then the number of non-isomorphic exceptional simple objects in* $\operatorname{mod}(X_{n+1})$ *is equal to* $n + |A_z| + |B_z| - O$.

The proof of this theorem will follow rather easily from our previous results. We start with the following lemma.

LEMMA 11.2.2. *Assume that* $\mathcal{A}$, $\mathcal{B}$ *are two abelian categories and* $\mathcal{C} \subset \mathcal{A}$, $\mathcal{D} \subset \mathcal{B}$ *are two abelian subcategories closed under subquotients. Assume that there are inverse equivalence* F, G *between* $D^b(\mathcal{A})$ *and* $D^b(\mathcal{B})$. *Assume furthermore that for all* i, H^iF *sends* $\mathcal{C}$ *to* $\mathcal{D}$ *and* H^iG *sends* $\mathcal{D}$ *to* $\mathcal{C}$. *Then the maps*

$$(11.11)\qquad \begin{aligned} \bar{F} &: \mathcal{C} \to \mathcal{D} : [C] \mapsto \sum (-)^i [H^i(FC)] \\ \bar{G} &: \mathcal{D} \to \mathcal{C} : [D] \mapsto \sum (-)^i [H^i(GD)] \end{aligned}$$

define isomorphisms between the Grothendieck groups of $\mathcal{C}$ *and* $\mathcal{D}$.

PROOF. Let $\bar{\mathcal{C}}$ and $\bar{\mathcal{D}}$ be the closures of $\mathcal{C}$ and $\mathcal{D}$ under extensions. Then clearly F and G define inverse equivalences between $D^b_{\bar{\mathcal{C}}}(\mathcal{A})$ and $D^b_{\bar{\mathcal{D}}}(\mathcal{B})$. Hence we obtain (using standard isomorphisms for Grothendieck groups)

$$K_0(\mathcal{C}) \cong K_0(\bar{\mathcal{C}}) \cong K_0(D^b_{\bar{\mathcal{C}}}(\mathcal{A})) \cong K_0(D^b_{\bar{\mathcal{D}}}(\mathcal{B})) \cong K_0(\bar{\mathcal{D}}) = K_0(\mathcal{D})$$

The composition of these isomorphisms (and their inverses) is given by (11.11). □

We deduce the following result.

LEMMA 11.2.3. *Let* $p_1, \dots, p_n \in Y$, $n \leq 8$ *be such that* $(\sigma p_i)_i$ *are in general position. Let* $z \in \mathbb{N}(Y/\langle\tau\rangle)$. *Then* $K_0(M_z(\tilde{X}_i))$ *and* $K_0(M_z(X_{i+1}))$ *are isomorphic for* $i \leq n$ *(see §6.9 for notations).*

PROOF. By Theorem 11.1.3 we know that $L\delta_i^*$ and $R\delta_{i,*}$ are inverse equivalences between $D^b(\tilde{X}_i)$ and $D^b(X_{i+1})$.

Now using the explicit construction of the functors T_p in Proposition 5.7.2 it is easy to verify that for $\mathcal{M} \in \operatorname{trans}_Y(X_{i+1})$ we have $T_p(\mathcal{M}) = T_p(\delta_i^*\mathcal{M})$. Similarly using Proposition 7.2.4 we have for $\mathcal{N} \in \operatorname{trans}_Y(\tilde{X})$ the identity $T_p(\delta_{i,*}\mathcal{N}) = T_p(\mathcal{N})$. Finally using Theorem 9.1.9 we find that the higher derived functors of δ_i^* and $\delta_{i,*}$ maps $\operatorname{mod}(X_{i+1})$ to $\operatorname{iso}_Y(\tilde{X}_i)$ and $\operatorname{mod}(\tilde{X}_i)$ to $\operatorname{iso}_Y(X_{i+1})$.

It follows that we can apply lemma 11.2.2 to obtain an isomorphism between $M_z(\tilde{X}_i)$ and $M_z(X_{i+1})$. □

We now obtain.

LEMMA 11.2.4. *Let* $p_1, \dots, p_n \in Y$, $n \leq 8$ *be such that* $(\sigma p_i)_i$ *are in general position. Let* z *and* O *be as in the statement of Theorem 11.2.1 (where we identify* $\mathbb{N}F$ *with* $\mathbb{N}(Y/\langle\tau\rangle)$*). Then the number of non-isomorphic exceptional simple objects in* $\operatorname{mod}(X_{n+1})$ *is equal to* $n + \operatorname{rk} K_0(M_z(X_0)) - O$.

PROOF. By iterating Theorem 6.9.1 and lemma 11.2.4 we easily find

$$\operatorname{rk} K_0(M_0(X_{n+1})) = n + \operatorname{rk} K_0(M_z(X_1)) - O$$

It is now easy to see that $M_0(X_{n+1})$ is equivalent to $\mathrm{iso}_Y(X_{n+1})$. Clearly the exceptional simple objects in $\mathrm{mod}(X_{n+1})$ coincide with the simple objects in $\mathrm{iso}_Y(X_{n+1})$. It now suffices to check that $\mathrm{iso}_Y(X_{n+1})$ is a finite length category.

Note that by construction $X_{n+1} = \operatorname{Proj} A_{n+1}$ where A_{n+1} is a noetherian graded ring containing a regular central element t in degree one (see §9.1 and in particular Prop. 9.1.3). If M is a graded A_{n+1}-module and $\mathcal{M} = \pi M$ then multiplication by t corresponds to the map $\mathcal{M}(-Y) \to \mathcal{M}$. It easily follows that if M is a graded A_{n+1}-module of Gelfand-Kirillov dimension > 1 then πM is not in $\mathrm{iso}_Y(X_{n+1})$. Hence objects in $\mathrm{iso}_Y(X_{n+1})$ come from graded A-modules of Gelfand-Kirillov dimension one. By considering multiplicity one finds that the latter form a finite length category when viewed in $\operatorname{Proj} A_{n+1}$. □

Proof of Theorem 11.2.1. By lemma 11.2.4 we have to compute the abelian group $K_0(M_z(X_0))$ in the special case that $|z| \leq 6$ (by our assumption that $n \leq 6$).

Remember that $X_0 = \operatorname{Proj} A$ where A is a three dimensional elliptic Artin-Schelter regular algebra. By considering the action of the central element in degree three it is easily seen that every object in $\mathrm{trans}_Y(X_0)$ is of the form πM with $\operatorname{GKdim} M \leq 2$. Since the graded A-modules with $\operatorname{GKdim} \leq 1$ correspond to objects in $\mathcal{C}_f$ [**8**] it follows by considering multiplicity that $\mathrm{trans}_Y(X_0)/\mathcal{C}_f$ is a finite length category. The same holds for $M_z(X_0)$ so a basis for $K_0(M_z(X_0))$ is given by the isomorphism classes of simple objects. These simple objects have been classified in Theorem 10.2.6. There are in 1-1 correspondence with the following two sets.

$$A'_z = \{D \in \mathbb{N}(Y/\langle\tau\rangle) \mid D \leq z, D \text{ is compatible with } L\}$$

$$B'_z = \{D \in \mathbb{N}(Y/\langle\tau\rangle) \mid D \leq z, D \text{ is compatible with } 2L$$
$$\text{and } D \text{ is not a sum } D_1 + D_2 \text{ with } D_i \text{ compatible with } L\}$$

where L represents the divisor of a line in $\mathbb{P}^2$. Thus $\operatorname{rk} K_0(M_z(X_{n+1})) = |A'_z| + |B'_z|$. It is now easy to see that A_z is in bijection with A'_z and similarly B_z is in bijection with B'_z. This finishes the proof. □

CHAPTER 12

Non-commutative cubic surfaces

In this chapter we recycle notations and assumptions from Chapter §11. We will however assume in addition that $n = 6$. Thus will fix six points $(p_j)_j$ on Y and our aim will be to study X_7. Since we will make no use of Theorem 11.1.3, we will not assume that the points (σp_j) are in general position. Furthermore we will also not assume that τ has infinite order.

By construction $X_7 = \operatorname{Proj} F$ for a certain graded k-algebra F. Since the hypotheses for Proposition 9.2.3 hold on X_6 (lemma 11.1.1) we find by that proposition that F contains a regular central element t in degree one such that $\bar{F} = F/tF$ is the twisted homogeneous coordinate ring [**10**] associated to the triple $(Y, \mathcal{N}_7, \tau)$. Here $\mathcal{N}_7$ is a line bundle of degree $9 - 6 = 3$, defined by (11.1).

From these data we can compute the Hilbert-series of F. We find $H(F, s) = (1 - s^3)/(1 - s)^4$. This suggests that X_7 should be viewed as a non-commutative cubic surface. In this chapter we substantiate this intuition by showing that there exists a 4-dimensional Artin-Schelter regular algebra P [**7**], containing a normal element C in degree three such that $F = P/(C)$. We can then view $\operatorname{Proj} P$ as a quantum $\mathbb{P}^3$ which contains X_7 as a cubic divisor.

REMARK 12.1. Of course the commutative analogue of this is well-known. If one blows up 6 points in general position in $\mathbb{P}^2$ then one obtains a cubic surface in $\mathbb{P}^4$ [**25**, Thm 24.4, p119]. However the reader may wonder why, in the non-commutative case, we don't need that our points are in general position. The explanation is of course that X_7 is not a straight blowing up of X_1, but is constructed by repeatedly applying the constructions $X_j \mapsto \tilde{X}_j \mapsto X_{j+1}$. In the commutative case, if the points are in general position, then δ_j is an isomorphism between $\tilde{X}_j$ and X_{j+1} and so in that case X_7 is indeed a straight blowing up of X_1. This will in general not be true in the non-commutative case, except in a derived sense. See Theorem 11.1.3.

The construction of P is easy. It follows from [**8**] that $\bar{F}$ has a (minimal) presentation

$$\bar{F} = k[x_1, x_2, x_3]/(r_1, r_2, r_3, C_3)$$

where $\deg x_1 = 1$, $\deg r_i = 2$, $\deg C_3 = 3$. One deduces that F has a presentation

$$F = k[x_1, x_2, x_3, t]/(r_1', r_2', r_3', C_3', [t, x_1], [t, x_2], [t, x_3])$$

where r_i', C_3' are homogeneous liftings of r_i, C_3.

We now put

$$P = k[x_1, x_2, x_3, t]/(r_1', r_2', r_3', [t, x_1], [t, x_2], [t, x_3])$$

and we will show that P is Artin-Schelter regular. To this end we make use of the fact that by [**8**] one knows that $\bar{P} = P/tP = k[x_1, x_2, x_3]/(r_1, r_2, r_3)$ is a three-generator three-dimensional Artin-Schelter regular algebra [**7**]. We can then use the criterion [**21**, Cor. 2.7].

To state this criterion we let R_P and $R_{\bar{P}}$ stand for the relations of degree two in P and $\bar{P}$. Suppose we have the following

1. Left and right multiplication by t is injective on P_1.
2. The image of $(P_1 \otimes R_P) \cap (R_P \otimes P_1)$ under the natural map $P^{\otimes 3} \to \bar{P}^{\otimes 3}$ is $(\bar{P}_1 \otimes R_{\bar{P}}) \cap (R_{\bar{P}} \otimes \bar{P}_1)$.

Then according to [**21**, Cor. 2.7], P will be Artin-Schelter regular with Hilbert series $1/(1-s)^4$.

We now verify conditions 1.,2. Condition 1. follows from the observations that $P_{\leq 2} = F_{\leq 2}$ and that F is a domain since $\bar{F}$ is a domain.

Hence we concentrate on condition 2. To simplify the notations we put $V = \sum_i kx_i$, $R_2 = \sum_i kr_i \subset V^{\otimes 2}$, $R_3 = kC_3 \subset V^{\otimes 3}$, $W = V \oplus kt$, $S_2 = (\sum_i kr'_i) + (\sum_i k[t, x_i]) \subset W^{\otimes 2}$, $S_3 = kC'_3 \subset W^{\otimes 3}$. With these notations

$$\begin{aligned} \bar{F} &= k[V]/(R_2 \oplus R_3) \\ F &= k[W]/(S_2 \oplus S_3) \\ P &= k[W]/(S_2) \\ \bar{P} &= k[V]/(R_2) \end{aligned}$$

Note that $R_P = S_2$, $R_{\bar{P}} = R_2$.

We first claim that the following complexes are exact in degrees ≤ 3.

$$0 \to (V \otimes R_2 \cap R_2 \otimes V) \otimes \bar{F} \to (R_2 \oplus R_3) \otimes \bar{F} \to V \otimes \bar{F} \to \bar{F} \to k \to 0 \tag{12.1}$$

$$0 \to (W \otimes S_2 \cap S_2 \otimes W) \otimes F \to (S_2 \oplus S_3) \otimes F \to W \otimes F \to F \to k \to 0 \tag{12.2}$$

Let us first consider (12.1). The only place where exactness is non-obvious is at $(R_2 \oplus R_3) \otimes \bar{F}$. Hence it is sufficient to show that the alternating sum of the Hilbert series of the terms in (12.1) is zero in degrees ≤ 3. This easily follows from the fact

$$\dim(V \otimes R_2 \cap R_2 \otimes V) = 1$$

which is true because $\bar{P}$ is Koszul.

We use the same method to check the exactness of (12.2). This time we need

$$\dim(W \otimes S_2 \cap S_2 \otimes W) = 4$$

Now we know that the dimensions of

$$\begin{aligned} F_2 &= W^{\otimes 2}/(S_2) \\ F_3 &= W^{\otimes 3}/(S_3 + W \otimes S_2 + S_2 \otimes W) \end{aligned}$$

are equal to 10 and 19 respectively. This yields that

$$\begin{aligned} \dim S_2 &= 16 - 10 = 6 \\ \dim(S_3 + W \otimes S_2 + S_2 \otimes W) &= 64 - 19 = 45 \end{aligned}$$

Since by [**8**] we have $R_3 \cap (V \otimes R_2 + R_2 \otimes V) = 0$, it also follows that $S_3 \cap (W \otimes S_2 + S_2 \otimes W) = 0$. Thus

$$\dim(W \otimes S_2 + S_2 \otimes W) = 44$$

Hence we obtain that

$$\dim(W \otimes S_2 \cap S_2 \otimes W) = 4 \times 6 + 6 \times 4 - 44 = 4$$

This proves what we want.

Now we tense (12.2) with $\bar{F}$ and we combine the result with (12.1) to form the following commutative diagram.

$$\begin{array}{ccccccccc}
 & & W \otimes \bar{F} & \xrightarrow{\gamma} & k \otimes \bar{F} & & & & \\
 & & \alpha\downarrow & & \beta\downarrow & & & & \\
(W \otimes S_2 \cap S_2 \cap W) \otimes \bar{F} & \longrightarrow & (S_2 \oplus S_3) \otimes \bar{F} & \longrightarrow & W \otimes \bar{F} & \longrightarrow & \bar{F} & \longrightarrow & k \\
\downarrow & & \downarrow & & \downarrow & & \downarrow & & \\
(V \otimes R_2 \cap R_2 \cap V) \otimes \bar{F} & \longrightarrow & (R_2 \oplus R_3) \otimes \bar{F} & \longrightarrow & V \otimes \bar{F} & \longrightarrow & \bar{F} & \longrightarrow & k \\
 & & \downarrow & & \downarrow & & & & \\
 & & 0 & & 0 & & & &
\end{array}$$

where in addition the rows start with monomorphisms and end with epimorphisms.

α, β and γ are defined by

$$\alpha(w \otimes 1) = (tw - wt) \otimes 1$$
$$\beta(1 \otimes 1) = t \otimes 1$$
$$\gamma(w \otimes 1) = 1 \otimes \bar{w}$$

The homology of the middle complex is given by $\operatorname{Ext}^i_F(k, \bar{F})$ in degrees ≤ 3. In particular this complex is exact at $(S_2 \oplus S_3) \otimes \bar{F}$.

It is also clear that γ is surjective in degrees ≥ 1. A trivial diagram chase now shows that

$$W \otimes S_2 \cap S_2 \otimes W \to V \otimes R_2 \cap R_2 \otimes V$$

is surjective. This completes the proof of conditions 1. and 2. above.

So at this point we know that P is Artin-Schelter regular with Hilbert series $1/(1-s)^4$. We still have to show that C'_3 is a regular normalizing element in P.

By looking at Hilbert series it is clear that t is a regular central element in P. Hence since $\bar{P}$ is a domain by [**8**], the same holds for P. In particular C'_3 is regular in P. Looking at Hilbert series of P and F reveals that the twosided ideal (C'_3) in P must be free of rank one on the left and on the right. Hence $(C'_3) = C'_3 P = P C'_3$ and thus C'_3 is normalizing. This completes the proof.

APPENDIX A

Two-categories

A 2-category is a category where the homsets themselves are categories. The objects of such a category are called 0-cells, the arrows are called 1-cells and the arrows between arrows are called 2-cells. Such 2-cells are drawn as follows

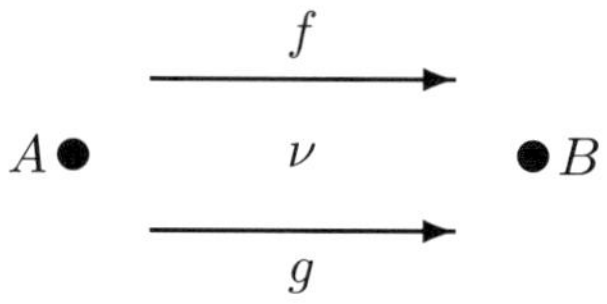

As usual arrows can be composed, and so can 2-cells. It turns out that 2-cells even have two compositions. Vertical ones

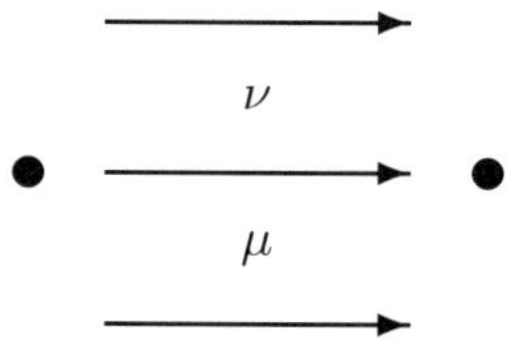

denoted by $\mu \cdot \nu$, which come from the composition in $\mathrm{Hom}(A, B)$ and horizontal ones

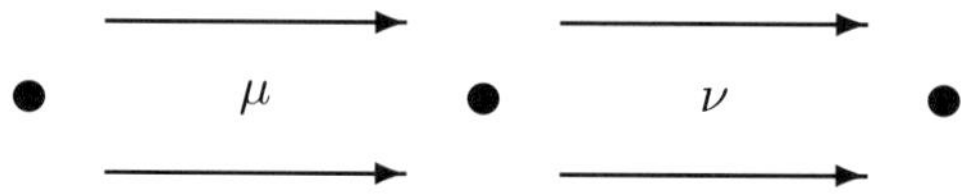

denoted by $\mu\nu$ which come from the fact that the pairing $\mathrm{Hom}(B, C) \times \mathrm{Hom}(A, B) \to \mathrm{Hom}(A, C)$ has to be a bifunctor. Between those two compositions there is a natural compatibility. Assume that one has the following diagram

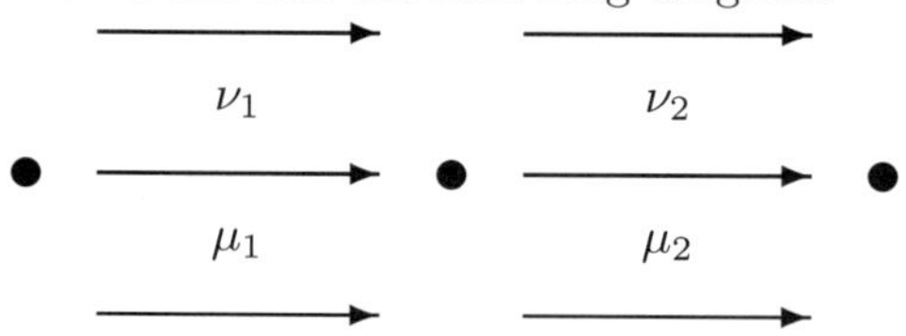

then one has $\mu_1\mu_2 \cdot \nu_1\nu_2 = (\mu_1 \cdot \nu_1)(\mu_2 \cdot \nu_2)$. Of course since a set is a category with only the identity arrows we can consider every category trivially as a 2-category.

The archetypal example of a 2-category is "**Cat**," the category of all categories. In this case the objects are the categories (living in some universe), the arrows are the functors and the 2-cells are the natural transformations. It is therefore not surprising that the standard properties of categories and functors can be mimicked inside a 2-category.

For example an arrow $A \xrightarrow{f} B$ is a left adjoint of an arrow $B \xrightarrow{g} A$ if there is a unit $\eta : \mathrm{id}_A \to gf$ and a counit $\epsilon : fg \to \mathrm{id}_B$ satisfying the standard associativity conditions. As usual g is determined by f up to unique isomorphism. If in this situation the η and ν are isomorphisms then we call f, g inverse equivalences and we say that A and B are equivalent.

In a 2-category it is natural not only to consider ordinary commutative diagrams (so-called "strict" commutative diagrams) but also pseudo-commutative diagrams. These diagrams are commutative up to *explicit* isomorphism. For example the notation

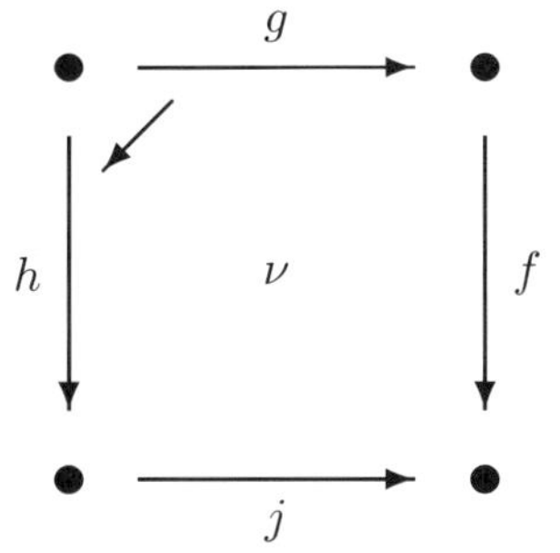

(A.1)

means that there is an isomorphism $fg \xrightarrow{\nu} jh$. Often ν is clear from the context. In such a case we sometimes tacitly ignore ν and treat (A.1) as a real commutative diagram.

The naturality of such pseudo-commutative diagrams is reflected in the definition of a pseudo-functor [**20**] between 2-categories, which we give below. Assume that $\mathcal{C}$, $\mathcal{D}$ are 2-categories. A pseudo-functor $T : \mathcal{C} \to \mathcal{D}$ associates to every object of $\mathcal{C}$ an object of $\mathcal{D}$, to every arrow $f : A \to B$ of $\mathcal{C}$ an arrow $T(f) : T(A) \to T(B)$ of $\mathcal{D}$ and to every 2-cell $\nu : f \to g$ a two-cell $T(\nu) : T(f) \to T(g)$. If T were an ordinary functor then we would require that for compositions of arrows fg one has $T(fg) = T(f)T(g)$. However for a pseudo-functor we only require the existence of isomorphisms $\eta_{f,g} : T(f)T(g) \to T(fg)$ which we consider as being part of the description of T. The data describing T has to satisfy a list of compatibilities which may be summarized by saying that every diagram that can commute must commute.

One can go on and define natural transformations between pseudo-functors and even natural transformations between natural transformations ("modifications"). In this way pseudo-functors between 2-categories form themselves a 2-category and the category of all 2-categories is a 3-category!

We will call a pseudo-functor $S : \mathcal{C} \to \mathcal{D}$ an *equivalence* if for every object D in $\mathcal{D}$ there exists an object C in $\mathcal{C}$ such that D is equivalent to $S(C)$ (essential surjectivity) and for all objects A, B in $\mathcal{C}$ the canonical map

$$\mathrm{Hom}_{\mathcal{C}}(A, B) \to \mathrm{Hom}_{\mathcal{D}}(SA, SB)$$

is an equivalence of categories. As usual $\mathcal{C}$ and $\mathcal{D}$ are said to be equivalent if there exists an equivalence $S : \mathcal{C} \to \mathcal{D}$. One verifies that such an S has a quasi-inverse (in an appropriate sense) and hence "equivalence of two-categories is symmetric".

An example of a pseudo-functor between 2-categories is given by adjunction. Assume that $\mathcal{C}$ is a 2-category in which every arrow f possesses a right adjoint Rf. Then R defines a pseudo-functor $R : \mathcal{C} \to \mathcal{C}^{\mathrm{opp}}$. If every arrow f has also a left adjoint Lf then R, L are inverse equivalences of 2-categories between $\mathcal{C}$ and $\mathcal{C}^{\mathrm{opp}}$.

One more bit of notation. If $\mathcal{C}$ is a 2-category and A is an object of $\mathcal{C}$ then the relative category $\mathcal{C}/A$ is the 2-category of pairs (B, f) where B is an object of $\mathcal{C}$ and $f : B \to A$ is an arrow. An arrow $(B, f) \to (C, g)$ in $\mathcal{C}/A$ is given by an arrow $h : B \to C$ together with an isomorphism $\mu : f \to gh$. A 2-cell $(h, \mu) \to (h', \mu')$ is given by an isomorphism $\nu : h \to h'$ such that $(\mathrm{id}_g \nu) \cdot \mu = \mu'$.

APPENDIX B

Summary of notations

Symbol	Meaning	Section
k	an algebraically closed field	§1
$\mathrm{Inj}(\mathcal{D})$	injective objects in $\mathcal{D}$	§3.1
$\mathcal{L}(\mathcal{D},\mathcal{C})$	left exact functors from $\mathcal{D}$ to $\mathcal{C}$	§3.1
$\mathrm{BIMOD}(\mathcal{C}-\mathcal{D})$	the opposite categorie of $\mathcal{L}(\mathcal{D},\mathcal{C})$	§3.1
$\mathrm{Bimod}(\mathcal{C}-\mathcal{D})$	objects in $\mathrm{BIMOD}(\mathcal{C}-\mathcal{D})$ having a left adjoint	§3.1
$\otimes$	composition of bimodules	§3.1
$\mathrm{MOD}(\mathcal{C})$	the category $\mathrm{BIMOD}(\mathbf{Ab}-\mathcal{C})$	§3.1
$\mathcal{H}om_{\mathcal{C}}(\mathcal{M},-)$	the left exact functor represented by $\mathcal{M}$	§3.1
$\mathcal{E}xt$	the derived functor of $\mathcal{H}om$	§3.1
$\mathcal{T}or$	a kind of derived functor of "$\otimes$"	§3.1
$\mathrm{ALG}(\mathcal{D})$	the algebra objects in $\mathrm{BIMOD}(\mathcal{D}-\mathcal{D})$	§3.1
$\mathrm{Alg}(\mathcal{D})$	the algebra objects in $\mathrm{Bimod}(\mathcal{D}-\mathcal{D})$	§3.1
$\mathrm{Mod}(\mathcal{A})$	the module category of the algebra $\mathcal{A}$	§3.1
$"\varprojlim"$	the virtual inverse limit	§3.1
$\mathrm{BIGR}(\mathcal{C}-\mathcal{D})$	graded "weak" $\mathcal{C}$-$\mathcal{D}$ bimodules	§3.2
$\mathrm{Bigr}(\mathcal{C}-\mathcal{D})$	graded $\mathcal{C}$-$\mathcal{D}$ bimodules	§3.2
$\mathrm{GRALG}(\mathcal{D})$	the algebra objects in $\mathrm{BIGR}(\mathcal{D}-\mathcal{D})$	§3.2
Gralg	the algebra objects in $\mathrm{Bigr}(\mathcal{D}-\mathcal{D})$	§3.2
$\mathrm{Gr}(\mathcal{A})$	the graded modules over the algebra $\mathcal{A}$	§3.2
$\mathcal{S}(\mathcal{A})$	a certain Serre subcategory of $\mathrm{Gr}(\mathcal{A})$	§3.10
i_*, i^*, $i^!$	functors associated to an inclusion i	§3.3
$\mathrm{Mod}(X)$	the category of objects associated to X	§3.5
$\mathcal{O}_X$	a distinguished object in X	§3.5
$\Gamma(X,-)$	notation for the functor $\mathrm{Hom}_X(\mathcal{O}_X,-)$	§3.5
o_X	the bimodule given by the identity functor	§3.5
$\mathrm{Bimod}(X)$	notation for $\mathrm{Bimod}(\mathrm{Mod}(X)-\mathrm{Mod}(X))$	§3.5
$\mathrm{Alg}(X)$	notation for $\mathrm{Alg}(\mathrm{Mod}(X)$	§3.5
Sch	a category of good schemes	§3.5
Qsch	the category of quasi-schemes	§3.5
Qsch/X	the category of quasi-schemes over X	§3.5
$\mathrm{Spec}\,\mathcal{A}$	a quasi-scheme with module category $\mathrm{Mod}(\mathcal{A})$	§3.5
$o_X(-Y)$	a subbimodule of o_X associated to a divisor Y	§3.6
$o_X(nY)$	notation for $o_X(-Y)^{\otimes -n}$	§3.6
$\mathcal{M}(nY)$	notation for $\mathcal{M}\otimes o_X(nY)$	§3.6
$\mathcal{N}_{Y/X}$	the "normal bundle" of Y in X	§3.6
$\mathrm{Tors}_Y(X)$, $\mathrm{Iso}_Y(X)$	certain categories associated to $Y\subset X$	§3.6
$\mathrm{Tors}(\mathcal{A})$	torsion modules over a graded algebra $\mathcal{A}$	§3.7
$\mathrm{QGr}(\mathcal{A})$	the category $\mathrm{Gr}(\mathcal{A})/\mathrm{Tors}(\mathcal{A})$	§3.7

τ	the "torsion functor"	§3.7
π	the quotient functor $\mathrm{Gr}(\mathcal{A}) \to \mathrm{QGr}(\mathcal{A})$	§3.7
ω	the right adjoint to π	§3.7
$(\tilde{-})$	the composition $\omega\pi$	§3.7
$\mathrm{Proj}\,\mathcal{A}$	A quasi-scheme whose category is $\mathrm{QGr}(\mathcal{A})$	§3.7
$\mathrm{Pqsch}\,/X$	"polarized" quasi-schemes over X	§3.7
$Q\mathcal{S}(\mathcal{A})$	the image of $\mathcal{S}(\mathcal{A})$ in $\mathrm{QGr}(\mathcal{A})$	§3.11
$\alpha^{-1}(\mathcal{S})$	an alternative notation for $Q\mathcal{S}(\mathcal{A})$	§3.11
$\mathcal{A}^{(n)}$	the n'th Veronese of $\mathcal{A}$	§3.12
$L_i\alpha^*$	a kind of derived functor to α^*	§3.9
$\mathrm{PC}(A)$	the category of pseudo-compact A-modules	§4
$\mathrm{Top}(A)$	the category of topological A-modules	§4
$\mathrm{Dis}(A)$	the category of discrete A-modules	§4
$\mathrm{PCFin}(A)$	the category of pseudo-compact f.l. modules	§4
$\mathrm{PC}(A-B)$	the category of pseudo-compact A-B-bimodules	§4
$\hat{\otimes}$	completed tensor product	§4
X	a quasi-scheme	§5.1
Y	a commutative curve which is a divisor in X	§5.1
p	a fixed point on Y	§5.1
τ	an automorphism of Y associated to $\mathcal{N}_{Y/X}$	§5.1
$\mathcal{N}_\tau$	the twist of the line bundle $\mathcal{N}$ by τ	§5.1
$O_\tau(p)$	the τ-orbit of p	§5.1
$\mathcal{O}_p$	the object in $\mathrm{Mod}(X)$ corresponding to $p \in Y$	§5.1
$\mathcal{C}_f$	finite length objects supported on Y	§5.1
$\mathcal{C}$	direct limits of objects in $\mathcal{C}_f$	§5.1
$\mathcal{C}_p$	objects in $\mathcal{C}$ supported on the τ-orbit of p	§5.1
$\mathcal{C}_{f,p}$	the finite length objects in $\mathcal{C}_p$	§5.1
C_p	the pseudo-compact ring associated to $\mathcal{C}_p$	§5.1
N	a canonical normal element in C_p	§5.1
$(\hat{-})_p$	the completion functor	§5.1ff
R	the completion of $\mathcal{O}_Y$ at p	§5.1
m	the maximal ideal of R	§5.1
U	the generator of the maximal ideal of R	§5.1
m_i	the i'th maximal ideal of C_p	§5.1
S_i	the i'th simple C_p-module	§5.1
e_i	the i'th diagonal primitive idempotent in C_p	§5.1
P_i	the i'th pseudo-compact projective of C_p	§5.1
o_q	the bimodule on X corresponding to $q \in Y$	§5.1
$\mathrm{cohBIMOD}(o_X - o_X)$	coherent bimodules on X	§5.5
$\tilde{\mathcal{C}}_{f,p}$	finite extensions of $o_{\tau^i p}$	§5.5
$\mathrm{trans}_Y(X)$	objects "transversal" to Y	§5.5
$\mathrm{Div}(\mathcal{F})$	the divisor on Y associated to $\mathcal{F}$	§5.5
$\mathcal{F}_Y$	notation for $\mathcal{F}/\mathcal{F}(-Y)$	§5.7
$T_p(\mathcal{F})$	an invariant for $\mathcal{F} \in \mathrm{trans}_Y(X)$	§5.7
$N_p(\mathcal{F})$	the p-normalization of $\mathcal{F}$	§5.7
m_q	the ideal in o_X associated to o_q	§6.1
$m_{Y,q}$	the ideal in o_Y associated to o_q	§6.1
I	notation for $m_p(Y)$	§6.1

I_Y	notation for $m_{Y,p}(Y)$	§6.1
μ	the multiplicity of p on Y	§6.1
$\mathcal{D}$	the Rees algebra for I	§6.2
$\mathcal{D}_Y$	the Rees algebra for I_Y	§6.2
$\tilde{X}$	the blowup of X at p	§6.3
$\tilde{Y}$	the strict transform of Y in $\tilde{X}$	§6.3
α, β, i, j	maps in a diagram relating X,$\tilde{X}$, Y,$\tilde{Y}$	§6.3
τ'	the lifting of τ to $\tilde{Y}$	§6.3
$\alpha^{-1}(\mathcal{C}_p)$	objects supported on the exceptional curve	§6.5
L	the exceptional curve	§6.6
$\mathrm{Mod}(L)$	the objects on L	§6.6
$\mathcal{O}_L$	the "structure sheaf" on L	§6.6
α^{-1}	the non-normalized strict transform	§6.8
α_s^{-1}	the normalized strict transform	§6.8
$l_p(\mathcal{F})$	the Loewy length of $T_p(\mathcal{F})$	§6.8
$M_z(X)$	A certain subcategory of $\mathrm{trans}_Y(X)/\mathcal{C}_f$	§6.9
$D^*(Z)$, $D^*_f(Z)$	derived categories of a quasi-scheme Z	§10
$(Y, \mathcal{L}, \sigma)$	a "triple" in the sense of [**8**]	§10
A	the regular algebra associated to $(Y, \mathcal{L}, \sigma)$	§10
g	the canonical central element in A	§10
$o_X(1)$	the bimodule given by the shift on $\mathrm{Gr}(A)$	§10
$r_{p,n}(\mathcal{M})$	the multiplicities of $\mathcal{M}$ infinitely near p	§10
$e(\mathcal{M})$	the multiplicity of $\mathcal{M}$	§10
$\overline{\mathrm{Div}}(\mathcal{M})$	a variant on $\mathrm{Div}(\mathcal{M})$	§10

APPENDIX C

Index of terminology

Term	Section
(weak) bimodule	§3.1
coherent bimodule	§3.1
virtual inverse limit	§3.1
torsion inverse system	§3.1
(weak) algebra	§3.1
graded notions	§3.2
(weakly) closed subcategory	§3.3
(weak) ideals	§3.4
well-closed	§3.4
very well-closed	§3.4
Rees algebra	§3.4
quasi-scheme	§3.5
enriched quasi-scheme	§3.5
Spec	§3.5
divisor	§3.6
normal bundle	§3.6
Proj	§3.7
condition "χ"	§3.8
pseudo-compact ring	§4
completion	§5.3, §5.4 and §5.6
multiplicity	§5.7
blowing up	§6
exceptional curve	§6.6
strict transform	§6.8
admissible maps and compositions	§7.2
semi-orthogonal decomposition	§8.1
lines and conics	§10.2
exceptional simple object	§11.2
two-category	App. A

Bibliography

[1] K. Ajitabh and M. Van den Bergh, *Presentation of critical modules of GK-dimension 2 over elliptic algebras*, Proc. Amer. Math. Soc. **127** (1999), no. 6, 1633–1639.

[2] K. Ajitabh, *Modules over regular algebras and quantum planes*, Ph.D. thesis, MIT, 1994.

[3] ———, *Modules over elliptic algebras and quantum planes*, Proc. London Math. Soc. (3) **72** (1996), 567–587.

[4] M. Artin, *Some problems on three-dimensional graded domains*, Representation theory and algebraic geometry (Waltham, MA, 1995) (Cambridge), Cambridge Univ. Press, Cambridge, 1997, pp. 1–19.

[5] M. Artin and J. T. Stafford, *Noncommutative graded domains with quadratic growth*, Invent. Math. **122** (1995), 231–276.

[6] M. Artin and J. J. Zhang, *Noncommutative projective schemes*, Adv. in Math. **109** (1994), no. 2, 228–287.

[7] M. Artin and W. Schelter, *Graded algebras of global dimension 3*, Adv. in Math. **66** (1987), 171–216.

[8] M. Artin, J. Tate, and M. Van den Bergh, *Some algebras associated to automorphisms of elliptic curves*, The Grothendieck Festschrift, vol. 1, Birkhäuser, 1990, pp. 33–85.

[9] ———, *Modules over regular algebras of dimension 3*, Invent. Math. **106** (1991), 335–388.

[10] M. Artin and M. Van den Bergh, *Twisted homogeneous coordinate rings*, J. Algebra **188** (1990), 249–271.

[11] A. Beilinson, *Coherent sheaves on $\mathbb{P}^n$ and problems of linear algebra*, Functional Anal. Appl. **12** (1978), 214–216.

[12] M. Böckstedt and A. Neeman, *Homotopy limits in triangulated categories*, Compositio Math. **86** (1993), 209–234.

[13] A. I. Bondal and A. E. Polishchuk, *Homological properties of associative algebras: the method of helices*, Russian Acad. Sci. Izv. Math **42** (1994), 219–260.

[14] A. I. Bondal, *Representations of associative algebras and coherent sheaves*, Math. USSR-Izv. **34** (1990), no. 1, 23–42.

[15] W. Fulton, *Algebraic curves*, W. A. Benjamin, Inc, New York, Amsterdam, 1969.

[16] P. Gabriel, *Des catégories abéliennes*, Bull. Soc. Math. France **90** (1962), 323–448.

[17] A. Grothendieck, *Sur quelques points d'algèbre homologiques*, Tôhoku Math. J. (2) **9** (1957), 119–221.

[18] R. Hartshorne, *Residues and duality*, Lecture notes in mathematics, vol. 20, Springer Verlag, Berlin, 1966.

[19] ———, *Algebraic geometry*, Springer-Verlag, 1977.

[20] G. M. Kelly and R. Street, *Review of the elements of 2-categories*, Category Seminar, vol. 420, Springer Verlag, 1972, pp. 75–103.

[21] L. Le Bruyn, S. P. Smith, and M. Van den Bergh, *Central extensions of three-dimensional Artin-Schelter regular algebras*, Math. Z. **222** (1996), 171–212.

[22] L. Le Bruyn, *Conformal $\mathfrak{sl}_2$ enveloping algebras*, Comm. Algebra **23** (1995), no. 4, 1325–1362.

[23] V. A. Lunts and A. L. Rosenberg, *Localization for quantum groups*, Selecta Math. (N.S.) **5** (1999), no. 1, 123–159.

[24] S. MacLane, *Categories for the working mathematician*, Springer Verlag, Berlin, 1971.

[25] Y. I. Manin, *Cubic forms: Algebra, geometry, arithmetic*, North Holland, Amsterdam, 1974.

[26] C. Nastacescu and F. Van Oystaeyen, *Graded ring theory*, North-Holland, 1982.

[27] D. O. Orlov, *Projective bundles, monoidal transformations and derived functors of coherent sheaves*, Russian Acad. Sci. Izv. Math **41** (1993), no. 1, 133–141.

[28] A. L. Rosenberg, *Non-commutative algebraic geometry and representations of quantized algebras*, Mathematics and its Applications, vol. 330, Kluwer Academic Publishers, Dordrecht, 1995.

[29] ———, *The spectrum of abelian categories and reconstruction of schemes*, Rings, Hopf algebras, and Brauer groups (Antwerp/Brussels, 1996) (New York), Dekker, New York, 1998, pp. 257–274.

[30] S. P. Smith and J. J. Zhang, *Curves on quasi-schemes*, Algebr. Represent. Theory **1** (1998), no. 4, 311–351.

[31] B. Stenström, *Rings of quotients*, Die Grundlehren der mathematischen Wissenschaften in Einzeldarstellungen, vol. 217, Springer Verlag, Berlin, 1975.

[32] R. Thomason and T. Trobaugh, *Higher algebraic K-theory of schemes and of derived categories*, The Grothendieck Festschrift, vol. 3, Birkhäuser, 1990, pp. 247–435.

[33] M. Van den Bergh, *The non-commutative Cremona transform*, in preparation.

[34] M. Van den Bergh, *A translation principle for Sklyanin algebras*, J. Algebra **184** (1996), 435–490.

[35] ———, *Abstract blowing down*, Proc. Amer. Math. Soc. **128** (2000), no. 2, 375–381.

[36] M. Van Gastel and M. Van den Bergh, *Graded modules of Gelfand-Kirillov dimension one over three-dimensional Artin-Schelter regular algebras*, J. Algebra **196** (1997), no. 1, 251–282.

[37] A. B. Verevkin, *On a non-commutative analogue of the category of coherent sheaves on a projective scheme*, Amer. Math. Soc. Transl. **151** (1992), 41–53.

[38] J. J. Zhang, *Twisted graded algebras and equivalences of graded categories*, Proc. London Math. Soc. (3) **72** (1996), no. 2, 281–311.

Editorial Information

To be published in the *Memoirs*, a paper must be correct, new, nontrivial, and significant. Further, it must be well written and of interest to a substantial number of mathematicians. Piecemeal results, such as an inconclusive step toward an unproved major theorem or a minor variation on a known result, are in general not acceptable for publication. Papers appearing in *Memoirs* are generally longer than those appearing in *Transactions*, which shares the same editorial committee.

As of May 31, 2001, the backlog for this journal was approximately 7 volumes. This estimate is the result of dividing the number of manuscripts for this journal in the Providence office that have not yet gone to the printer on the above date by the average number of monographs per volume over the previous twelve months, reduced by the number of volumes published in four months (the time necessary for preparing a volume for the printer). (There are 6 volumes per year, each containing at least 4 numbers.)

A Consent to Publish and Copyright Agreement is required before a paper will be published in the *Memoirs*. After a paper is accepted for publication, the Providence office will send a Consent to Publish and Copyright Agreement to all authors of the paper. By submitting a paper to the *Memoirs*, authors certify that the results have not been submitted to nor are they under consideration for publication by another journal, conference proceedings, or similar publication.

Information for Authors

Memoirs are printed from camera copy fully prepared by the author. This means that the finished book will look exactly like the copy submitted.

The paper must contain a *descriptive title* and an *abstract* that summarizes the article in language suitable for workers in the general field (algebra, analysis, etc.). The *descriptive title* should be short, but informative; useless or vague phrases such as "some remarks about" or "concerning" should be avoided. The *abstract* should be at least one complete sentence, and at most 300 words. Included with the footnotes to the paper should be the 2000 *Mathematics Subject Classification* representing the primary and secondary subjects of the article. The classifications are accessible from `www.ams.org/msc/`. The list of classifications is also available in print starting with the 1999 annual index of *Mathematical Reviews*. The Mathematics Subject Classification footnote may be followed by a list of *key words and phrases* describing the subject matter of the article and taken from it. Journal abbreviations used in bibliographies are listed in the latest *Mathematical Reviews* annual index. The series abbreviations are also accessible from `www.ams.org/publications/`. To help in preparing and verifying references, the AMS offers MR Lookup, a Reference Tool for Linking, at `www.ams.org/mrlookup/`. When the manuscript is submitted, authors should supply the editor with electronic addresses if available. These will be printed after the postal address at the end of the article.

Electronically prepared manuscripts. The AMS encourages electronically prepared manuscripts, with a strong preference for $\mathcal{A}_{\mathcal{M}}\mathcal{S}$-LaTeX. To this end, the Society has prepared $\mathcal{A}_{\mathcal{M}}\mathcal{S}$-LaTeX author packages for each AMS publication. Author packages include instructions for preparing electronic manuscripts, the *AMS Author Handbook*, samples, and a style file that generates the particular design specifications of that publication series. Though $\mathcal{A}_{\mathcal{M}}\mathcal{S}$-LaTeX is the highly preferred format of TeX, author packages are also available in $\mathcal{A}_{\mathcal{M}}\mathcal{S}$-TeX.

Authors may retrieve an author package from e-MATH starting from `www.ams.org/tex/` or via FTP to `ftp.ams.org` (login as `anonymous`, enter username as password, and type `cd pub/author-info`). The *AMS Author Handbook* and the *Instruction Manual* are available in PDF format following the author packages link from `www.ams.org/tex/`. The author package can be obtained free of charge by sending email to `pub@ams.org` (Internet) or from the Publication Division, American Mathematical Society, P.O. Box 6248, Providence, RI 02940-6248. When requesting an author package, please specify AMS-LaTeX or AMS-TeX, Macintosh or IBM (3.5) format, and the publication in which your paper will appear. Please be sure to include your complete mailing address.

Sending electronic files. After acceptance, the source file(s) should be sent to the Providence office (this includes any TeX source file, any graphics files, and the DVI or PostScript file).

Before sending the source file, be sure you have proofread your paper carefully. The files you send must be the EXACT files used to generate the proof copy that was accepted for publication. For all publications, authors are required to send a printed copy of their paper, which exactly matches the copy approved for publication, along with any graphics that will appear in the paper.

TeX files may be submitted by email, FTP, or on diskette. The DVI file(s) and PostScript files should be submitted only by FTP or on diskette unless they are encoded properly to submit through email. (DVI files are binary and PostScript files tend to be very large.)

Electronically prepared manuscripts can be sent via email to `pub-submit@ams.org` (Internet). The subject line of the message should include the publication code to identify it as a Memoir. TeX source files, DVI files, and PostScript files can be transferred over the Internet by FTP to the Internet node `e-math.ams.org` (130.44.1.100).

Electronic graphics. Comprehensive instructions on preparing graphics are available at `www.ams.org/jourhtml/graphics.html`. A few of the major requirements are given here.

Submit files for graphics as EPS (Encapsulated PostScript) files. This includes graphics originated via a graphics application as well as scanned photographs or other computer-generated images. If this is not possible, TIFF files are acceptable as long as they can be opened in Adobe Photoshop or Illustrator. No matter what method was used to produce the graphic, it is necessary to provide a paper copy to the AMS.

Authors using graphics packages for the creation of electronic art should also avoid the use of any lines thinner than 0.5 points in width. Many graphics packages allow the user to specify a "hairline" for a very thin line. Hairlines often look acceptable when proofed on a typical laser printer. However, when produced on a high-resolution laser imagesetter, hairlines become nearly invisible and will be lost entirely in the final printing process.

Screens should be set to values between 15% and 85%. Screens which fall outside of this range are too light or too dark to print correctly. Variations of screens within a graphic should be no less than 10%.

Inquiries. Any inquiries concerning a paper that has been accepted for publication should be sent directly to the Electronic Prepress Department, American Mathematical Society, P. O. Box 6248, Providence, RI 02940-6248.

Editors

This journal is designed particularly for long research papers, normally at least 80 pages in length, and groups of cognate papers in pure and applied mathematics. Papers intended for publication in the *Memoirs* should be addressed to one of the following editors. In principle the Memoirs welcomes electronic submissions, and some of the editors, those whose names appear below with an asterisk (*), have indicated that they prefer them. However, editors reserve the right to request hard copies after papers have been submitted electronically. Authors are advised to make preliminary email inquiries to editors about whether they are likely to be able to handle submissions in a particular electronic form.

Algebra to CHARLES CURTIS, Department of Mathematics, University of Oregon, Eugene, OR 97403-1222 email: `cwc@darkwing.uoregon.edu`

Algebraic geometry and commutative algebra to LAWRENCE EIN, Department of Mathematics, University of Illinois, 851 S. Morgan (M/C 249), Chicago, IL 60607-7045; email: `ein@uic.edu`

Algebraic topology and cohomology of groups to STEWART PRIDDY, Department of Mathematics, Northwestern University, 2033 Sheridan Road, Evanston, IL 60208-2730; email: `priddy@math.nwu.edu`

Combinatorics and Lie theory to SERGEY FOMIN, Department of Mathematics, University of Michigan, Ann Arbor, Michigan 48109-1109; email: `fomin@math.lsa.umich.edu`

Complex analysis and complex geometry to DUONG H. PHONG, Department of Mathematics, Columbia University, 2990 Broadway, New York, NY 10027-0029; email: `phong@math.columbia.edu`

***Differential geometry and global analysis** to LISA C. JEFFREY, Department of Mathematics, University of Toronto, 100 St. George St., Toronto, ON Canada M5S 3G3; email: `jeffrey@math.toronto.edu`

***Dynamical systems and ergodic theory** to ROBERT F. WILLIAMS, Department of Mathematics, University of Texas, Austin, Texas 78712-1082; email: `bob@math.utexas.edu`

Functional analysis and operator algebras to BRUCE E. BLACKADAR, Department of Mathematics, University of Nevada, Reno, NV 89557; email: `bruceb@math.unr.edu`

Geometric topology, knot theory and hyperbolic geometry to ABIGAIL A. THOMPSON, Department of Mathematics, University of California, Davis, Davis, CA 95616-5224; email: `thompson@math.ucdavis.edu`

Harmonic analysis, representation theory, and Lie theory to ROBERT J. STANTON, Department of Mathematics, The Ohio State University, 231 West 18th Avenue, Columbus, OH 43210-1174; email: `stanton@math.ohio-state.edu`

***Logic** to THEODORE SLAMAN, Department of Mathematics, University of California, Berkeley, CA 94720-3840; email: `slaman@math.berkeley.edu`

Number theory to MICHAEL J. LARSEN, Department of Mathematics, Indiana University, Bloomington, IN 47405; email: `larsen@math.indiana.edu`

***Ordinary differential equations, partial differential equations, and applied mathematics** to PETER W. BATES, Department of Mathematics, Brigham Young University, 292 TMCB, Provo, UT 84602-1001; email: `peter@math.byu.edu`

***Partial differential equations and applied mathematics** to BARBARA LEE KEYFITZ, Department of Mathematics, University of Houston, 4800 Calhoun Road, Houston, TX 77204-3476; email: `keyfitz@uh.edu`

***Probability and statistics** to KRZYSZTOF BURDZY, Department of Mathematics, University of Washington, Box 354350, Seattle, Washington 98195-4350; email: `burdzy@math.washington.edu`

***Real and harmonic analysis and geometric partial differential equations** to WILLIAM BECKNER, Department of Mathematics, University of Texas, Austin, TX 78712-1082; email: `beckner@math.utexas.edu`

All other communications to the editors should be addressed to the Managing Editor, WILLIAM BECKNER, Department of Mathematics, University of Texas, Austin, TX 78712-1082; email: `beckner@math.utexas.edu`.

Selected Titles in This Series

(Continued from the front of this publication)

703 **Erik Guentner, Nigel Higson, and Jody Trout,** Equivariant E-theory for C^*-algebras, 2000

702 **Ilijas Farah,** Analytic guotients: Theory of liftings for quotients over analytic ideals on the integers, 2000

701 **Paul Selick and Jie Wu,** On natural coalgebra decompositions of tensor algebras and loop suspensions, 2000

700 **Vicente Cortés,** A new construction of homogeneous quaternionic manifolds and related geometric structures, 2000

699 **Alexander Fel′shtyn,** Dynamical zeta functions, Nielsen theory and Reidemeister torsion, 2000

698 **Andrew R. Kustin,** Complexes associated to two vectors and a rectangular matrix, 2000

697 **Deguang Han and David R. Larson,** Frames, bases and group representations, 2000

696 **Donald J. Estep, Mats G. Larson, and Roy D. Williams,** Estimating the error of numerical solutions of systems of reaction-diffusion equations, 2000

695 **Vitaly Bergelson and Randall McCutcheon,** An ergodic IP polynomial Szemerédi theorem, 2000

694 **Alberto Bressan, Graziano Crasta, and Benedetto Piccoli,** Well-posedness of the Cauchy problem for $n \times n$ systems of conservation laws, 2000

693 **Doug Pickrell,** Invariant measures for unitary groups associated to Kac-Moody Lie algebras, 2000

692 **Mara D. Neusel,** Inverse invariant theory and Steenrod operations, 2000

691 **Bruce Hughes and Stratos Prassidis,** Control and relaxation over the circle, 2000

690 **Robert Rumely, Chi Fong Lau, and Robert Varley,** Existence of the sectional capacity, 2000

689 **M. A. Dickmann and F. Miraglia,** Special groups: Boolean-theoretic methods in the theory of quadratic forms, 2000

688 **Piotr Hajłasz and Pekka Koskela,** Sobolev met Poincaré, 2000

687 **Guy David and Stephen Semmes,** Uniform rectifiability and quasiminimizing sets of arbitrary codimension, 2000

686 **L. Gaunce Lewis, Jr.,** Splitting theorems for certain equivariant spectra, 2000

685 **Jean-Luc Joly, Guy Metivier, and Jeffrey Rauch,** Caustics for dissipative semilinear oscillations, 2000

684 **Harvey I. Blau, Bangteng Xu, Z. Arad, E. Fisman, V. Miloslavsky, and M. Muzychuk,** Homogeneous integral table algebras of degree three: A trilogy, 2000

683 **Serge Bouc,** Non-additive exact functors and tensor induction for Mackey functors, 2000

682 **Martin Majewski,** ational homotopical models and uniqueness, 2000

681 **David P. Blecher, Paul S. Muhly, and Vern I. Paulsen,** Categories of operator modules (Morita equivalence and projective modules, 2000

680 **Joachim Zacharias,** Continuous tensor products and Arveson's spectral C^*-algebras, 2000

679 **Y. A. Abramovich and A. K. Kitover,** Inverses of disjointness preserving operators, 2000

678 **Wilhelm Stannat,** The theory of generalized Dirichlet forms and its applications in analysis and stochastics, 1999

677 **Volodymyr V. Lyubashenko,** Squared Hopf algebras, 1999